T0291636

CAMBRIDGE MONOGRAPHS
ON MECHANICS AND APPLIED MATHEMATICS

GENERAL EDITORS

G. K. BATCHELOR, PH.D., F.R.S.
Lecturer in Mathematics in the University of Cambridge

H. BONDI, M.A.
Professor of Applied Mathematics at King's College,
University of London

COSMIC ELECTRODYNAMICS

COSMIC ELECTRODYNAMICS

BY

J. W. DUNGEY, PH.D.

Lecturer in Mathematics at King's College, Newcastle upon Tyne

CAMBRIDGE
AT THE UNIVERSITY PRESS
1958

CAMBRIDGE
UNIVERSITY PRESS

University Printing House, Cambridge CB2 8BS, United Kingdom

Cambridge University Press is part of the University of Cambridge.

It furthers the University's mission by disseminating knowledge in the pursuit of education, learning and research at the highest international levels of excellence.

www.cambridge.org
Information on this title: www.cambridge.org/9781316601877

First published 1958
First paperback edition 2015

A catalogue record for this publication is available from the British Library

ISBN 978-1-316-60187-7 Paperback

CONTENTS

4 Magnetostatics

5 Magnetodynamics

6 Acceleration of Charged Particles to High Energy

7 Solar Phenomena

PREFACE

In this monograph I have attempted to present a systematic account of the subject and to include a section on each major topic. The physics of the subject is discussed and the mathematical methods are explained, but few extended pieces of mathematics are presented; in particular only the results of Chandrasekhar's extensive work on problems of stability are quoted. The subject was developing rapidly during the period of writing, as may be seen from the dates in the bibliography. I have tried to bring my text up to date to August 1956, but excluding work reported at the symposium held by the I.A.U. in Stockholm in that month (Lehnert, 1958).

Before starting to write I enjoyed many stimulating discussions on various branches of cosmic electrodynamics with Dr R. G. Giovanelli. I am very grateful to Professor Bondi for his detailed and practical editing of more than half of the first draft and to Dr R. Latham for helpful comments on it. An I.C.I. Fellowship at Cambridge has provided ideal circumstances for writing.

I am also grateful to Dr R. Hide and Dr P. H. Roberts for their careful proof-reading and for the valuable suggestions which they made.

J. W. D.

CHAPTER 1

PRELIMINARIES

1.1. Definition of the subject

Astrophysics, and particularly solar physics, has reached a stage where many of the most interesting problems are believed to involve electromagnetism. It is by now accepted that electromagnetic fields play an essential role in nearly all transient features of the sun. It is possible, even, that they are important factors in stellar evolution, and an electromagnetic theory of the origin of the planets has been proposed (Alfvén, 1954). At the same time, the subject of cosmic electrodynamics is still in an early stage of development; the theoretical side of the subject is rather abstract and its results apply to models, whose connexion with observations is generally very tenuous. This book is intended primarily to explain which physical effects are likely to be important, and to present the simplest and most fundamental of those theoretical results, which are expected to be useful, without attempting to provide a final explanation of any particular observations. Some detailed theories have been proposed to account for various phenomena but, in the present state of the subject, it may be more desirable to try to consolidate the foundations than to build an elaborate superstructure.

As this book is primarily theoretical, and good accounts of the observations are available, an exposition of the theory is given first. This is followed by a brief discussion of the observations in the light of the theory, with occasional attempts to sketch the kind of theory which appears to be needed. This discussion also demonstrates the utility of the theoretical results obtained in the earlier chapters.

Before starting on theory, the scope of this work must be defined. Electromagnetic phenomena on the sun, magnetic storms and aurorae provide the bulk of the problems. The theoretical foundations laid in the early chapters can also be used to discuss stellar and interstellar fields, but these are treated more briefly. The mechanisms available for the acceleration of cosmic rays are discussed, but the other aspects of the problem of the origin of cosmic rays are only outlined. Radioastronomy is excluded from this work, as it is a separate though related subject. Some electrodynamic problems of the ionosphere, though they differ considerably from the other problems, are discussed in the last chapter, but problems of the earth's interior are not included.

One recent development holds great promise for the future of solar observations. This is Babcock's instrument, the solar magnetograph, which has been operating since 1952; a similar instrument is now operating in Cambridge. This instrument measures the Zeeman effect with a very sensitive technique and can detect fields of only a fraction of gauss. It scans the solar disc and yields a plot of the line-of-sight component of the magnetic field against position. Several very interesting results have already been obtained, such as:

(i) a 'general' field of about 2 gauss has been discovered at high latitudes;

(ii) sunspots are found to form in bipolar magnetic regions of much greater area and weaker fields than spots themselves;

(iii) an M-region has been found to coincide with a unipolar magnetic region;

(iv) it is found that there is frequently a weak irregular field and a preliminary study of its rate of variation with time has been made.

It is important to realize that the area of the sun from which light enters the instrument at any setting is considerably greater than that of even a large sunspot, and that smaller features cannot be seen (sunspots cause a deflexion because their fields are so strong). It should be possible, however, to construct a similar instrument with substantially greater resolving power. It may be confidently expected, then, that many new discoveries will come from this instrument during the next few solar cycles and that an instrument with greater resolving power will be built and will further increase our knowledge of solar magnetism. With this stimulus it is unlikely that the theoretical side of the subject will be allowed to stagnate.

Some laboratory experiments have been performed, which exhibit the same physical effects as those important in cosmic electrodynamics, and so can help in the elucidation of the cosmic problems. One kind of experiment requires a liquid with a very large electrical conductivity σ; if the scale length for the magnetic field is a, the decay time of the current is $4\pi\sigma a^2/c^2$, which for cosmic systems is generally very large compared with the time scale of the phenomena. In the laboratory a is necessarily very much smaller, and it is found that no liquid has a large enough conductivity to make a correctly scaled model of a cosmic system. Liquid sodium has a higher conductivity than mercury, but it is awkward to handle; both have been used. With mercury it is quite difficult to investigate the required phenomena owing to the rapidity of their decay, but Alfvén waves have been studied and the experiments have provided confirmation of the theory. The inhibition by a magnetic field of turbulence in pipes and convection has also been studied.

Another kind of experiment was first performed by Birkeland in 1901, the terrella experiment, in which a magnetic dipole is placed near the path of a beam of electrons. The experiment is very beautiful and can be used to determine the orbits of cosmic-ray particles assuming a dipole field for the earth. To investigate magnetic storms, however, a neutral ionized beam would be needed and it is more difficult to obtain a neutral beam.

Certain difficulties face any attempt to construct laboratory models of astrophysical systems. One is the necessity for solid-bounding walls; the boundaries often have important and complicated effects and many phenomena in discharge tubes are still not well understood. On the electrodynamic side the fact that air is an insulator and the difficulty of achieving a high degree of ionization in a discharge tube are also important. Since the classical theory is already based on laboratory experiments, so that any discrepancy between it and new experiments would imply a new discovery in laboratory electrodynamics, these experiments may be regarded as an alternative to the theoretical method of attack. While any experimental check on the theory is obviously welcome, the theoretical method seems easier, except perhaps for problems of turbulence.

1.2. Introduction to the theoretical approach

The fundamentals on which the theoretical work is based are classical and well established on the laboratory scale. It is obviously necessary to work out such a classical theory before considering possible modifications of the fundamental laws, which might be important only for phenomena of cosmic dimensions; for no new hypothesis can be checked by observation, unless the consequences of the classical theory have been worked out. For this reason suggestions such as that made by Schuster and Wilson, and later Blackett (1947), that a magnetic field is associated with the rotation of material, will not be discussed and Maxwell's equations will be assumed to be true.

The major part of the theoretical work involves a subject known as hydromagnetics which has only recently been developed although it is based on classical laws. Hydromagnetics is the study of the behaviour of conducting fluids in electromagnetic fields and is an extension of hydrodynamics and aerodynamics. The difficulties met are similar to those of aerodynamics; hydromagnetics is somewhat more complicated and there is a greater variety in the kinds of phenomena which can occur. One kind of difficulty arises from the non-linearity of the equations; much of the work that has been done has involved a linearized treatment of small disturbances.

Another difficulty, which occurs also in meteorology, is due to the fact that the observed phenomena have no simple symmetry; it is generally impracticable to set up models, which approximate at all closely to them. Mathematical models are useful, rather, to illustrate particular physical effects and to provide an evaluation of the relative importance of different effects. The understanding of simple models therefore assists in the qualitative understanding of the observations. At the same time it is important to remember that models are usually restricted by some sort of symmetry and, that consequently, there is a danger of thinking in only two dimensions, particularly when diagrams are used. The existence of the third dimension can radically alter a physical situation, particularly in regard to stability; a two-dimensional tight-rope walker confined to the vertical plane containing the rope could never fall off.

It is important to realize that the physical systems of cosmic electrodynamics are very different from those of laboratory electrodynamics. In cosmic electrodynamics it is usually necessary to contemplate strongly ionized gas stretching to infinity; this is unfamiliar because we live in a bubble of almost unionized gas. The kind of boundary condition required is very different from those of laboratory electrodynamics, and most cosmic systems differ from laboratory systems in that:

(i) the linear dimensions are very large;

(ii) there are no rigid bodies;

(iii) there are no electrical insulators.

Obvious exceptions are the planets for (ii) and (iii) and their atmospheres for (iii). The statement (iii) depends on the definition of an insulator, and it will be seen later that the maximum electron density for a gas to behave as an insulator decreases as the scale of the system increases. Some numerical orders of magnitude are given in § 1.3. No instance of the gas behaving as an insulator will be met until chapter 9, though it is possible that in the interior of a dark dust cloud, where starlight cannot penetrate, the electron density might be sufficiently small. The interstellar gas generally behaves as a conductor and it follows that there may be electric currents anywhere.

The solar atmosphere is a good electrical conductor and this has important consequences. It is natural, for instance, to picture a pair of sunspots as the poles of a horseshoe magnet, but the magnetic field in the solar atmosphere above the pair is not likely to be even approximately the same as the field of a horseshoe magnet in air. The latter field is determined by the absence of electric currents in the air. But the solar atmosphere is conducting and the current density is determined by mechanical considerations. The magnetic field in a sunspot is accompanied by variations in pressure and temperature, which must result in variations in pressure in the atmosphere above. If these pressure variations were not balanced by an electromagnetic force, the resulting motion would in general induce currents. This kind of problem is considered in chapter 4, but we may already reject the assumption that the current density in stellar atmospheres or interstellar space is negligible. This assumption has sometimes been used to justify the idea that cosmic fields approximate to dipole fields far from their sources. The same argument shows why cosmic electrodynamics is more akin to aerodynamics than to the theory of electromagnetic fields *in vacuo*. The problems do not involve the calculation of the field due to known sources; indeed it will be seen in § 1.4 that the sources are calculated from the field.

Consider next the generation of electric fields. Gaussian units will be used throughout so that the electric field **E** and magnetic field **H** have the same dimensions. The departures from unity of both the dielectric constant and the magnetic permeability are negligible. In defining **E** and **H** it is of course necessary to specify the frame of reference. Consider the generation of an electric field by dynamo action. A cosmic dynamo must be constructed entirely of ionized gas and must be immersed in ionized gas; because of (iii), there is a short circuit between every pair of points. Speaking in terms of circuits, inductance is the most important circuit element, because of (i) and (iii). The importance of inductance pervades nearly all the theoretical work, but it is more conveniently expressed, not in terms of circuits, but as the 'freezing' of the magnetic field into the material. This idea will be explained in chapter 3 and will be used extensively.

Faraday's law states that the rate of change of flux through a closed loop is equal to the e.m.f. round the loop. For a loop of perfectly conducting material the e.m.f. must be zero and hence the flux cannot change. If the perfectly conducting material is fluid, this result can be extended to apply to a loop, every point of which moves with the velocity **u** of the fluid at that point. The condition of perfect conductivity requires that at each point the electric field in the frame moving with the local velocity **u** should vanish, or

$$\mathbf{E} = -\mathbf{u} \wedge \mathbf{H}/c. \qquad (1.1)$$

It is shown in chapter 2 that (1.1) is nearly always a good approximation for cosmic systems. A special application of a classical theorem of hydrodynamics shows that (1.1) implies that the flux through any closed loop moving with the material is constant*, which is a precise statement of the 'freezing' of the field into the material; a different proof is given in §3.2.

The approximate validity of (1.1) will be used in the next section to justify some important approximations in a general way. It should be noted that, if (1.1) is true in one frame, it is true in all frames; it is in fact a relativistically invariant equation. It may also be noted that, since the electric field is greatly dependent on the

* See M. Abraham & R. Becker: *The Classical Theory of Electricity and Magnetism*, 2nd ed., Blackie, 1950, p. 39.

frame of reference while the magnetic field is not, it is usual to think of the magnetic field as the primary feature; for instance, the existence of a magnetic field is independent of the frame.

It may be asked whether the idea of the 'freezing' of the field is applicable to the problem of a rotating laboratory magnet. This is more complicated than the cosmic problem, since it depends on the properties of the material from which the magnet is made and also differs in that the magnet is surrounded by insulating air. It may also be noted that if rotating axes are used, the fundamental equations are not the same as for non-rotating axes. The rotation of magnetic stars is discussed in §5.3.

1.3. Orders of magnitude

In any work of this kind it is necessary to make rough calculations of orders of magnitude using a characteristic length a and a characteristic time t to obtain the orders of magnitude of derivatives; e.g. the order of magnitude of $|\text{curl}\,\mathbf{H}|$ is taken to be H/a, a letter in ordinary italics denoting the magnitude of the corresponding vector. The reader should ponder a little the validity of this method, however; obviously the values of $|\text{curl}\,\mathbf{H}|$ at different points may differ by large factors, and hence need not all be even of the order of magnitude H/a. It is important to note that while values much less than H/a may well occur over large regions of space, values much greater than H/a can occur only in regions with one dimension correspondingly small compared with a. Such regions are sometimes important; an example will occur in §6.3. In using this method it is also necessary to consider whether there are two or more characteristic lengths differing by a considerable factor; this may occur when there is turbulence and more care is then needed.

Another relation that is often roughly true is $ut \sim a$. The weaker assumption $ct > a$ is sufficient for some purposes, however.

Before considering numerical values we can demonstrate the negligibility of two terms in the fundamental equations in a general way, using the approximate validity of (1.1), which gives

$$E \sim uH/c,$$

and using $$u \ll c,$$

and $$a < ct.$$

Consider the Maxwell equation

$$4\pi\mathbf{j}=c\,\mathrm{curl}\,\mathbf{H}-\partial\mathbf{E}/\partial t, \tag{1.2}$$

we have $$|\mathrm{curl}\,\mathbf{H}|\sim H/a,$$

and $$|\partial\mathbf{E}/\partial t|\sim uH/ct;$$

hence $$|\partial\mathbf{E}/\partial t|\ll c\,|\mathrm{curl}\,\mathbf{H}|,$$

and $$4\pi\mathbf{j}\approx c\,\mathrm{curl}\,\mathbf{H}. \tag{1.3}$$

This approximation, which is the neglect of the displacement current, separates the theory of cosmic electrodynamics from that of radioastronomy. The general order of magnitude argument shows that, if a solution of a cosmic electrodynamic problem is obtained by neglecting the displacement current, this neglect can be justified *a posteriori*, since the solution is consistent with it. The problem of the stability of the solution has received little attention as such; it has been studied as a mechanism for the spontaneous generation of radio noise, though with rather inconclusive results. If the solution were unstable, some sort of high-frequency oscillations or noise would be generated and would react on the slowly varying part of the solution. This possibility remains to be investigated, but is not expected to be important, except perhaps in shock waves. Because of the rapid temporal variations involved in a shock wave, the neglect of the displacement current may not be valid for a shock wave in a magnetic field. If it were not valid, radio noise would be generated and it is clearly important that shock waves should be studied without making this approximation; but this is difficult and has not yet been done.

A further general inequality is obtained from the Maxwell equation $4\pi\rho=\mathrm{div}\,\mathbf{E}$.

Because $$|\mathbf{j}\wedge\mathbf{H}|/c\sim H^2/4\pi a,$$

and $$\rho E\sim u^2H^2/4\pi ac^2,$$

we have $$\rho E\ll|\mathbf{j}\wedge\mathbf{H}|/c.$$

Consequently, the force density due to space charge is negligible compared with that due to currents. Similarly, the convection current density $\rho\mathbf{u}$ is of the order of $u^2H^2/4\pi ac$ and is negligible compared with the total current density; in practice a stronger and more useful condition is valid, namely that the charge density of negative ions is very nearly equal to that of positive ions. This is shown in Table 1 of rough values, in which e denotes the magnitude of the electronic charge:

TABLE 1

	Chromosphere above sunspots	Corona	H II regions in interstellar space
a (cm.)	10^8	10^9	10^{11}
u (cm./sec.)	10^6	10^7	10^6
H (gauss)	10^3	10^2	10^{-5}
n (electrons/c.c.)	10^{12}	10^7	1
$\rho/ne \sim uH/4\pi neac$	5.10^{-14}	5.10^{-10}	5.10^{-13}
$j/ne \sim cH/4\pi nea$ (cm./sec.)	50	5.10^4	5.10^2

Though some of these values may be in error by considerable factors, the following conclusions are not in doubt. The penultimate row gives the ratio of the difference between the densities of positive and negative ions to the electron density n; the values of this ratio in Table 1 are seen to be very small, so that the densities of positives and negatives are very nearly equal. Combining this with the negligibility of its contribution $\rho\mathbf{E}$ to the force density, we see that ρ can be taken as zero in these cosmic systems. It should be stressed, however, that it would be incorrect to write $\operatorname{div}\mathbf{E}=0$; it has been assumed here that $\operatorname{div}\mathbf{E} \sim E/a$ and this has been used to show that ρ is small in certain particular senses. The last row of Table 1 gives the difference between the average velocities of the positive ions and electrons, calculated on the assumption that $j \sim cH/4\pi a$. The velocity difference so calculated is seen to be quite small; it is in fact considerably smaller than the thermal velocity of protons. This is significant because the thermal motion of the electrons controls the collision frequency between electrons and protons, so long as the thermal velocity of the electrons is greater than the difference between the average velocities of the electrons and protons. Since the values in Table 1 correspond to $j \sim cH/4\pi a$, velocity differences of these orders of magnitude correspond to sufficient currents to be important sources of the magnetic field, which shows that the gas does not behave as an insulator, as pointed out in the last section. The reverse argument is important in connexion with accelerating mechanisms (§6.3): if charged particles are accelerated to high energies by an electric field in some given region, all the charged particles in this region must be accelerated and then the current density must be vastly greater than $cH/4\pi a$. These conclusions are usually true for cosmic systems; in the photosphere n is larger than in the chromosphere; in H I regions in the interstellar gas, n is smaller than in H II regions, but probably by only a few powers of ten.

It is convenient to add here an approximation which is to be made in the Ohm's law equation, when it is derived in §2.7; it is this equation which leads to the approximate validity of equation (1.1) already mentioned. The equation is obtained by calculating $\partial \mathbf{j}/\partial t$ from the forces acting on the particles and will here be written

$$\frac{m_e}{ne^2}\frac{\partial \mathbf{j}}{\partial t} = \mathbf{E} + \mathbf{u}\wedge\mathbf{H}/c - \mathbf{j}/\sigma \quad \text{and other terms,} \tag{1.4}$$

where m_e is the electron mass, and it will be shown in §3.4 that the 'other terms' are small. Now from Maxwell's equations (1.2) and

$$\partial \mathbf{H}/\partial t = -c\,\mathrm{curl}\,\mathbf{E}, \tag{1.5}$$

we obtain $$4\pi\,\partial \mathbf{j}/\partial t = -c^2\,\mathrm{curl}\,\mathrm{curl}\,\mathbf{E} - \partial^2\mathbf{E}/\partial t^2,$$

whence $$|\,\partial \mathbf{j}/\partial t\,| \sim c^2E/4\pi a^2.$$

Then the condition for the left-hand side of (1.4) to be negligible is

$$a^2 \gg c^2 m_e/4\pi ne^2 \approx 3\cdot 10^{11} n^{-1}\ \text{cm.}^2 \tag{1.6}$$

The frequency $(4\pi ne^2/m_e)^{\frac{1}{2}}$ is known as the 'plasma frequency' and the quantity on the right-hand side of (1.6) may be called the square of the 'plasma wavelength'. The numerical values show that in cosmic systems this is always very small compared to the characteristic length a, and consequently that the left-hand side of (1.4) is negligible, so long as the method of obtaining orders of magnitude by means of the scale length is valid. It should be remembered, however, that Ohm's law is not valid, if the current density varies too rapidly with time.

1.4. Causal relationships

When reading the later chapters, the reader may feel the need for answers to the questions 'what drives the current?' and 'where does the electric field come from?', and this section is intended to anticipate such questions. In laboratory electrodynamics we are accustomed to think in terms of the answers: the current is driven by an electric field (often an induced field) and an electric field is produced by space charge and by a changing magnetic field. These answers may be included in a causal scheme for laboratory electrodynamics as follows

$$\begin{array}{ccccccc} & & & \rho & & & \mathbf{u} \\ & & \swarrow & & \nwarrow & \swarrow & \\ \dfrac{\partial \mathbf{H}}{\partial t} & \to & \mathbf{E} & \to & \mathbf{j} & \to & \mathbf{H}, \end{array}$$

where an arrow denotes 'causes'.

This section is intended to emphasize that the very familiar approach to electrodynamics, expressed in this scheme, is inappropriate to cosmic electrodynamics, and to warn the reader that a quite different approach will be used in later chapters. In fact, when ρ is omitted as being of no interest, the approach to be used for most of the problems reverses the above scheme. An account will now be given of how this scheme can be deduced from the equations involved. It is assumed that a solution of the equations is required satisfying given initial conditions; this is usually the case in cosmic problems.

Reverting to the general case, suppose that initial values are given for all the variables and it is desired to compute their values a short time later. In programming the computation one would find that the value of each variable is obtained from one particular equation and it would be necessary to work out which. Those equations which contain a time-derivative must determine that time-derivative and the remaining equations then determine the remaining variables in a unique way. Consequently, if a problem is to be discussed in terms of cause and effect, each variable must be the effect of the other terms in the equation determining that variable. When the causal scheme is defined in this way, it can therefore be altered by the neglect of various terms, particularly time-derivatives.

Consider now the basic equations of cosmic electrodynamics, which are:

(i) Maxwell's equations (1.3) and (1.5);

(ii) a form of Ohm's law to be derived in §2.7; it is equation (1.4) and usually reduces to (1.1);

(iii) the equation of transfer of momentum, determining $\partial\mathbf{u}/\partial t$, and involving the electromagnetic force density $\rho\mathbf{E}+\mathbf{j}\wedge\mathbf{H}/c$;

(iv) the equation of conservation of mass;

(v) the equation determining the rate of change of temperature.

First note that (1.5) requires $\mathbf{E}\rightarrow\mathbf{H}$ for any problem in which the initial values are given. Then (1.3) requires $\mathbf{H}\rightarrow\mathbf{j}$, but before discussing the scheme following from this, the method of deducing the scheme may be checked by considering the case, when the displacement current is not neglected. Then (1.2) must be used and requires $\mathbf{j}\rightarrow\mathbf{E}$ and $\mathbf{H}\rightarrow\mathbf{E}$. The two relations $\mathbf{E}\rightleftarrows\mathbf{H}$ are appropriate to electromagnetic radiation in free space and this makes sense when it is remembered how the concept of displacement current led Maxwell to predict radio propagation. The relation $\mathbf{j}\rightarrow\mathbf{E}$ would be appropriate for plasma oscillations.

Returning to cosmic electrodynamics and neglecting the displacement current, consider next the case where there is no motion of the gas. In (1.4) the term in $\partial\mathbf{j}/\partial t$ is neglected, and this equation determines $\mathbf{E}$ so that the scheme is

$$\begin{array}{c}\mathbf{E}\rightarrow\mathbf{H}\\ \nwarrow\ \swarrow\\ \mathbf{j}\end{array}.$$

This is appropriate to the decay of a magnetic field in static material, but most problems of cosmic electrodynamics involve motion. It may be helpful to state now that the most important effects are then represented by

$$\begin{array}{ccc}\mathbf{u}&\rightarrow&\mathbf{E}\\ \uparrow&&\downarrow\\ \mathbf{j}&\leftarrow&\mathbf{H}\end{array}$$

and that this may conveniently be simplified by eliminating $\mathbf{E}$ by means of (1.1) and $\mathbf{j}$ by means of (1.3) to give

$$\mathbf{u}\rightleftarrows\mathbf{H},$$

which is appropriate to Alfvén waves.

1.5. Orbits of particles

The calculation of the orbits of individual particles is not as important as might be thought at first sight, because the number of particles is so large that statistical methods are both justified and necessary. Such a calculation is possible only if the field is known, and, when the field depends on the orbits, a very difficult self-consistent field problem ensues. Consequently, the orbits of interest are usually those of minority particles, such as cosmic-ray particles, though it is doubtful whether even their effect on the field is always negligible. Extensive work has been done on orbit calculations, but only the simpler results will be given here.

In a uniform magnetic field the orbit of a charged particle is a helix with its axis parallel to the magnetic field. The angular gyro-frequency Ω is equal to eH/mc, where m is the mass of the particle and the radius of the spiral r is $v_\perp/\Omega$, where $v_\perp$ is the component of the particle's velocity perpendicular to the magnetic field. If the energy of the particle is relativistic, m must be taken to be the relativistic mass $m_0(1-v^2/c^2)^{-\frac{1}{2}}$, where v is the total velocity of the particle. The value of r for thermal protons is roughly $T^{\frac{1}{2}}H^{-1}$ cm.

If there is a uniform electric field perpendicular to the magnetic field and, if $E<H$, a charged particle moves in a trochoidal orbit and its average velocity is $\mathbf{u}$ given by (1.1) and is thus independent of the charge and mass of the particle. A gravitational field can be treated as equivalent to an electric field, but the equivalent electric field is proportional to m/e, and so is different for different kinds of particles.

In a magnetic field which is nearly uniform over distances of order r, the orbit still approximates to a helix and higher approximations can be obtained by treating the non-uniformity as a perturbation (Alfvén, 1950*a*). Two useful examples are worked out by the perturbation method in chapter 6.

The magnetic field due to a single charged particle moving in a circular orbit may be taken as that due to a current in a circular wire, and this approximates to a dipole field at large distances. The direction of rotation of a charged particle in a magnetic field is such that inside the orbit the magnetic field of the particle opposes the external field. For this reason an ionized gas may be regarded as being diamagnetic, but this approach does not prove fruitful, because it becomes awkward as soon as non-uniformities in either the gas or the field are considered. It may here be noted that any effects due to the actual dipole moments of the various particles in the gas are negligible.

The orbits in one particular kind of non-uniform field will now be described in order to illustrate the sort of fallacy that is apt to result when a problem is discussed entirely in terms of the orbits of the particles. Consider a magnetic field everywhere directed into the paper, but increasing in strength towards the top of the page. The orbit of a particle will have the shape shown in fig. 1.1,

Fig. 1.1

the curvature being greater where the field is stronger; the direction of motion depends on the sign of the charge. Thus positive and negative particles drift in opposite directions, and at first sight it is tempting to conclude that there is a current in the direction in which the positive ions drift. But the field due to a current in this direction would make the field strength decrease towards the top of the page, in contradiction to the original assumption. In other words, $\mathbf{j}$ and curl $\mathbf{H}$ would have opposite directions and (1.3) would be quite invalid; $\partial\mathbf{E}/\partial t$ would not be small, the field could not be in a steady state and in practice high-frequency oscillations would occur.

The conclusion that the current must always have the direction of the drift velocity of the positive ions is not in fact correct. Although each positive ion drifts to the right, the average velocity of the positive ions at a given point in space may be directed to the left because the average depends on the density of particles in the different orbits. If the density decreases towards the top of the page there will be more particles in the lower orbit of fig. 1.1 than in the upper orbit, so that the current may flow to the left. Similarly, if the temperature of the gas decreases towards the top of the paper, the magnitude of the velocity of a particle in the lower orbit will exceed that of a particle in the upper orbit, with the same possible result. In fact the condition of mechanical equilibrium in this sort of non-uniform magnetic field requires the gas pressure to decrease towards the top of the paper, and this explains qualitatively how (1.3) can be satisfied. It is clear that statistical methods are required for this type of problem, and this is one reason for using the Boltzmann theory of non-uniform gases as a basis; this is the subject of the next chapter.

The foregoing example illustrates the possibility of confusion, if not actual error, when the discussion is based entirely on the orbits of individual particles. It also shows that there is a danger that equation (1.3) will be forgotten, and some other fallacious method of calculating the current density substituted. This danger probably arises from our unfamiliarity with the idea that the magnetic field causes the current, as found in the last section. This causal relationship results from the neglect of the displacement current. It is not impossible that there should be a current density which does not satisfy (1.3), but, if there were, $\partial \mathbf{E}/\partial t$ would be large, radio-frequency oscillations would occur and these would so complicate the problem that, in all but trivial models, it would be unworkable. It is therefore easier to work out the slowly varying solutions, even at the expense of readjusting our ideas of the causal relationships.

The dynamics of an ionized gas would reduce to the calculation of orbits, if all the particles in a small volume had the same velocity; that is, if the thermal or other random velocities were negligible. This is seldom the case, but it has been assumed to be true by Chapman and Ferraro in their theory of magnetic storms; in their case it is justified because they are concerned with a beam of particles in a region where their source subtends a small solid angle, and because collisions are infrequent. Generally, the thermal motions are important in that the gas pressure is important. We now turn to the statistical method, which takes this into account.

CHAPTER 2

THE VELOCITY-DISTRIBUTION METHOD

2.1. Need for a statistical treatment

Cosmic phenomena evidently involve an enormous number of particles, and, since it is not practicable to calculate the orbit of every individual particle (in addition to the trouble explained in § 1.5), this fact necessitates the use of some statistical method. The use of statistics is in fact always justified by the number of individuals involved being very large, and it should be emphasized that the validity of such a method, when applied to a gas, does not depend on the frequency of collisions; it is the assumption of thermal conditions which depends on frequent collisions for its justification. A suitable statistical approach is provided by the Boltzmann velocity-distribution method, an account of which is given in this chapter. The gas is described by distribution functions, which are densities in phase space, that is the six-dimensional space having three co-ordinates of position and three of velocity (velocity is slightly more convenient than momentum). As there are six dimensions, the statistical description will be adequate, if the total number of particles is the sixth power of a large enough number; the values of na^3 given by Table 1 are 10^{36} for sunspots, 10^{34} for the Corona and 10^{33} for the interstellar gas, all of which are easily large enough. A further test is to consider the random (thermal) velocity of an element of the gas which must of course have dimensions much smaller than a, say $10^{-6}a$. The ratio of the random velocity of such an element to that of a single molecule is then $(10^{18}/na^3)^{\frac{1}{2}}$, which again is very small, and, as velocities very much less than that of sound are not important in cosmic phenomena, the method is adequate.

The inadequacy of other methods has been seen in § 1.5. It is tempting to discuss the orbits of a few particles and obtain conclusions concerning the whole system by regarding these few orbits as typical. This does not give a complete description of the system, however, since the relative densities of particles in different orbits are not discussed. Consequently the discussion of orbits is not a satisfactory substitute for the Boltzmann method, and, even when a knowledge of the orbits of particular particles is required, the Boltzmann method is still needed to determine the electromagnetic field.

In the Boltzmann theory the system is described by distribution functions for each type of molecule or ion in the gas, and average quantities appear which can be interpreted as describing continuous material. Small-scale variations are smoothed out in this statistical description, which of course provides less information than the exact description given by the position and velocity of each particle. The electromagnetic field used with the statistical description is also smoothed out; the average quantities obtained from the distribution functions determine charge and current densities, which are the sources of the smoothed field. The part of the field which has been smoothed away must not be neglected, but is treated separately; it represents the Coulomb interaction between neighbouring charged particles and radiation of short wavelengths. The effect of the Coulomb interaction between neighbouring particles is equivalent to the effect of collisions between the particles, and, if the ionization is appreciable (see § 3.4), this interaction is in fact more important than collisions between ions and neutral atoms.

2.2. Conditions peculiar to cosmic electrodynamics

An admirable and extensive account of the theory of non-uniform gases has been given by Chapman and Cowling (1939), which, together with Cowling's (1945) calculation of the conductivity of an ionized gas in a magnetic field, includes the results needed in the present work. Chapman and Cowling's work is very general, however, and for cosmic electrodynamic applications the accuracy of their results is greater than is needed in the present state of the subject. It therefore seems worth while to give a brief account of this work, simplified in a way which will suit the applications to be made later. The object of this account will be to explain the physical processes which occur, and to discuss the conditions of validity of the results. The most important simplification is obtained by using only a crude method of treating collisions; a great deal of complicated work is then avoided. In the applications it will be sufficient to know only the order of magnitude of the effects of collisions, but, should accurate values be needed, they are available in Chapman and Cowling (1939).

In cosmic electrodynamics the effects of collisions are not of first importance; this is another contrast with laboratory electrodynamics and requires a change of emphasis. The weakness of the effects of collisions does not necessarily imply that the conditions are non-thermal (meaning not even approximately thermal), but in some important applications they are. In this case the Chapman–Enskog method is not valid, and we therefore give in §2.9 some qualitative discussion of this difficult problem. An extreme example of non-thermal conditions is obtained by regarding cosmic-ray particles as part of the interstellar gas; though they are only a minute proportion of the total number of particles, their contribution to the stress tensor is an appreciable fraction of the total stress tensor. Note again that effects of this sort cannot be treated by orbit calculations.

2.3. Equations of motion for one constituent

In the Boltzmann theory the exact description of the gas in terms of the position $\mathbf{x}$ and velocity $\mathbf{v}$ of each particle is replaced by the statistical description in terms of distribution functions. The six-dimensional space whose co-ordinates are $\mathbf{x}$ and $\mathbf{v}$ is called 'phase-space' and the distribution functions are densities in phase space; they will be denoted by f_i, the suffix referring to the type of particle and being omitted in this section. Although the concept of a distribution function is just like that of density in ordinary space, its nature is worth some thought. For a useful description f must be almost constant over regions large enough to contain many particles; more precisely the number of particles in a volume V of phase space should not differ much from fV, where f is the value at the centre of the volume, provided V is large enough to make $fV \gg 1$ and small enough for f to be nearly constant over the whole volume V. Evidently the value of f at any point in phase space must depend on the positions in phase space of many particles. The definition of f is never precise. Since the total number of particles is very large, the volume V is small and it is permissible to take the fundamental property of f to be that the probable number of particles in the volume element $\mathbf{dx}\,\mathbf{dv}$ of phase space is $f\,\mathbf{dx}\,\mathbf{dv}$

(using the abbreviated notation $\mathbf{dx} = dx_1 \, dx_2 \, dx_3$). It is necessary for obvious physical reasons that $f \to 0$ as $v \to \infty$; in fact since the total energy must be finite, $fv^2 \to 0$.

The motion in phase space of each particle is determined by the equations

$$\frac{\mathbf{dx}}{dt} = \mathbf{v}, \tag{2.1}$$

$$\frac{\mathbf{dv}}{dt} = \frac{\mathbf{F}}{m}, \tag{2.2}$$

where the force $\mathbf{F}$ comes from the electric, magnetic and gravitational fields and so depends on $\mathbf{x}$ and $\mathbf{v}$. Only the smoothed parts of the fields are taken into account in $\mathbf{F}$ and these must be almost constant over regions large enough to contain many particles. Consequently, if the continuous distribution is thought of as a fluid in phase space, its velocity of flow is given sufficiently accurately by (2.1) and (2.2). Then f obeys the continuity equation

$$\frac{\partial f}{\partial t} = -\frac{\partial}{\partial x_i}(fv_i) - \frac{\partial}{\partial v_i}(fF_i/m) + \delta f + \delta' f, \tag{2.3}$$

where δf and $\delta' f$ represent the effects of collisions with particles of the same constituent and of other constituents respectively. A further term, not shown in (2.3), is needed to include the effects of ionization and recombination.

The number density n of particles in ordinary space is obviously given by

$$n = \int_{-\infty}^{\infty} f \, \mathbf{dv}, \tag{2.4}$$

where a single integral sign denotes integration over the whole of velocity space. The mean velocity of the particles at a point in ordinary space is $\mathbf{u}$ defined by

$$n\mathbf{u} = \int_{-\infty}^{\infty} \mathbf{v} f \, \mathbf{dv}. \tag{2.5}$$

Integration of (2.3) over the whole of velocity space yields an expression for $\partial n/\partial t$. The integral of the second term on the right-hand side vanishes because $f \to 0$ as $v \to \infty$. The collision terms do not contribute since there is no sudden change of the position of a particle, when a collision occurs. The result is the continuity equation for a continuous fluid

$$\frac{\partial n}{\partial t} = -\operatorname{div}(n\mathbf{u}). \tag{2.6}$$

By multiplying (2.3) by $m\mathbf{v}$ before integrating over velocity space, the macroscopic force density is obtained. Since momentum is conserved in collisions, there is no contribution from δf (this term represents collisions between pairs of like particles, both of which are included in the same distribution function). Then, transforming the second term on the right-hand side by integration by parts, the result is

$$m\frac{\partial}{\partial t}(nu_i) = -\frac{\partial T_{ij}}{\partial x_j} + \int_{-\infty}^{\infty} F_i f\,d\mathbf{v} + m\int_{-\infty}^{\infty} v_i \delta' f\,d\mathbf{v}, \tag{2.7}$$

with

$$T_{ij} = m\int_{-\infty}^{\infty} v_i v_j f\,d\mathbf{v}.$$

The mechanical stress tensor is $-S_{ij}$ defined by

$$S_{ij} = m\int_{-\infty}^{\infty} (v_i - u_i)(v_j - u_j) f\,d\mathbf{v}. \tag{2.8}$$

Using (2.6), (2.7) can then be written in the form

$$nm\left(\frac{\partial u_i}{\partial t} + u_j\frac{\partial u_i}{\partial x_j}\right) = -\frac{\partial S_{ij}}{\partial x_j} + \int_{-\infty}^{\infty} F_i f\,d\mathbf{v} + m\int_{-\infty}^{\infty} v_i \delta' f\,d\mathbf{u}, \tag{2.9}$$

which can be identified with the equation of transfer of momentum for a continuous fluid. For a thermal distribution S_{ij} reduces to $p\delta_{ij}$, where p is the partial pressure of the constituent described by f. The tensor S_{ij} is always symmetrical and in fact reduces to the form $p\delta_{ij}$ whenever the distribution function is isotropic.

One other useful result is obtained by multiplying (2.3) by $\frac{1}{2}mv^2$ before integrating over $\mathbf{v}$. If the temperature T is defined as $S_{kk}/3nk$ this result determines the rate of heating. Proceeding in a similar way to the derivation of (2.9) and using (2.6) and (2.9), we obtain

$$\tfrac{3}{2}nk\left(\frac{\partial T}{\partial t} + u_i\frac{\partial T}{\partial x_i}\right) = -\frac{\partial Q_i}{\partial x_i} - S_{ij}\frac{\partial u_j}{\partial x_i} + \tfrac{1}{2}m\int_{-\infty}^{\infty} v_i(v_i - 2u_i)\,\delta' f\,d\mathbf{v}, \tag{2.10}$$

where

$$Q_i = \tfrac{1}{2}m\int_{-\infty}^{\infty} (v_i - u_i)\,|\,\mathbf{v} - \mathbf{u}\,|^2 f\,d\mathbf{v}.$$

Q_i represents the flow of heat due to thermal conduction. It should be noted that **F** does not contribute directly to the heating, though it affects the relative velocities of different constituents and these control $\delta' f$.

2.4. Treatment of the electromagnetic field

The electromagnetic field to be substituted into **F** is to be a smoothed out field; it is to be calculated from Maxwell's equations with the values of ρ and **j** obtained from the distribution functions and its smoothness then follows from the smoothness of f. The actual field is not smooth, however, and therefore has to be divided into the smooth part defined in the way just described and the remaining part. It is convenient further to subdivide these two parts into two parts each, according to the following scheme:

(1) The smooth parts; the variation of this part is very small in distances $\sim n^{-\frac{1}{3}}$;

(1 *a*) the part whose variation is small in times $\sim (m_e/4\pi ne^2)^{\frac{1}{2}}$ (the 'plasma period') and in distances $\sim (m_e c^2/4\pi ne^2)^{\frac{1}{2}}$ (the 'plasma wavelength');

(1 *b*) the remainder of (1).

(2) The remainder when (1) is removed from the total field;

(2 *a*) the Coulomb field of particles within the smoothing distance;

(2 *b*) radiation of wavelength not very must greater than $n^{-\frac{1}{3}}$.

This analysis of the field is heuristic and not precise; it may be doubted whether the sum of (1), (2 *a*) and (2 *b*) does give the total field, but no other component has been suggested. Note that the Coulomb field of particles beyond the smoothing distance is accounted for in (1). The part (1 *b*) includes radio waves, but will be omitted completely in the present work. This omission is necessary in order to justify the neglect of the displacement current and another approximation, and, as remarked in § 1.3, this is where our way parts from that of theoretical radioastronomy. Radio waves do exert a steady force on an ionized gas, but this has not been found to be important. The part (2 *b*) also exerts radiation pressure, but will also be neglected; the radiation pressure on electrons can be obtained from the theory of the Compton effect, and is not found to be important. The effect of the part (2 *a*) can be represented as collisions between charged particles and is discussed next.

2.5. Collisions

The simplest method of treating collisions, the mean free-path method, regards each collision as occurring instantaneously; this is far from the case for the Coulomb interaction. Cowling has shown, however, that the effect of the Coulomb interaction is equivalent to the effect of collisions with a collision cross-section which is proportional to T^{-2} and depends logarithmically on the density of charged particles; an explanation of this result is given in the appendix. The expression for the collision frequency in terms of the collision cross-section A is well known. The average frequency, for particles of mass m_1, of collisions with particles of mass m_2 is ν_{12} given by

$$\nu_{12} = 2n_2 A[2kT(m_1+m_2)/\pi m_1 m_2]^{\frac{1}{2}}. \tag{2.11}$$

In the first approximation A is the same for any pair of charged particles and then, neglecting m_e compared to m_p, the collision frequencies for electrically neutral totally ionized hydrogen have the ratios

$$\nu_{ee} : \nu_{pp} : \nu_{ep} :: \sqrt{2} : (2m_e/m_p)^{\frac{1}{2}} : 1,$$

where the suffices e and p refer to electrons and protons.

A qualitative account of the effect of collisions will now be given; the general effect is to convert a systematic velocity distribution into a more random distribution. For instance, in an undisturbed enclosed gas, collisions make the distribution functions tend to the well-known Maxwellian form. It may be noted that the collisions representing the Coulomb interaction are elastic to a high order of approximation, so that conservation of the kinetic energy may be assumed. Collisions between like particles contribute to the change towards a Maxwellian distribution. The most important result is the change of S_{ij} towards greater isotropy and this may be expressed as a contribution to $\partial S_{ij}/\partial t$ of $\alpha\nu(\frac{1}{3}S_{kk}\delta_{ij} - S_{ij})$, where $\alpha \sim 1$ (for rigid spherical particles $\alpha = \frac{4}{5}$ and its value for the Coulomb interaction is quoted in §2.8). Elastic collisions between like particles do not change S_{kk}, because it is just twice the energy density of the random motions. The effect here described qualitatively is responsible for the attenuation of sound waves and also controls the coefficient of viscosity (see §2.8).

Collisions between unlike particles have a more important effect in addition to that already described and to their tendency to equalize the energies of random motion of the different particles; they tend to equalize the mean velocities of the two constituents involved, momentum being conserved. This process may be represented by a contribution to $\partial \mathbf{u}_1/\partial t$ of $\nu_{12}m_2(\mathbf{u}_2-\mathbf{u}_1)/(m_1+m_2)$. It has here been assumed that the direction of the relative velocity between the colliding particles is completely random after the collision, whereas for rigid spherical particles the direction tends to persist, so that the relative velocity after a collision has an average value in the direction it has before the collision; it is convenient to allow for this by adjusting the collision cross-section and to define the effective cross-section for the Coulomb interaction by this criterion. Collisions between unlike particles determine the velocities of diffusion in general, and in particular are responsible for the electrical resistivity of an ionized gas. It should be noted that, if the velocity of diffusion is comparable to the thermal velocity, the collision frequency depends on the velocity of diffusion, and the problem becomes much more difficult.

2.6. Equations of motion for totally ionized hydrogen

To describe totally ionized hydrogen, which is a good approximation to the composition of the gas in many cosmic applications, we need only two distribution functions, one for protons and one for electrons. Quantities referring to protons and electrons will be denoted by + and − superscripts except for the charge which will be written simply as $\pm e$, e being positive. The gravitational field will be omitted; its effect is easily seen. The force on a particle is then

$$\mathbf{F}^{\pm} = \pm e(\mathbf{E}+\mathbf{v}^{\pm}\wedge \mathbf{H}/c). \tag{2.12}$$

The collision terms will be omitted in this section and quoted from Cowling (1945) later; then (2.7) can be written, using (2.12),

$$m^{\pm}\frac{\partial}{\partial t}(n^{\pm}u_i^{\pm}) = -\frac{\partial}{\partial x_j}T_{ij}^{\pm} \pm n^{\pm}e(\mathbf{E}+\mathbf{u}^{\pm}\wedge \mathbf{H}/c)_i. \tag{2.13}$$

The magneto-ionic theory of the propagation of radio waves through an ionized gas is based on equation (2.13) for electrons, the motion of the positive ions being neglected. In cosmic electrodynamics, however, it is more convenient to rearrange the two equations (2.13) in terms of the following quantities, each of which has a simple physical significance:

the mass density $\mu = n^+m^+ + n^-m^-$,
the charge density $\rho = e(n^+ - n^-)$,
the current density $\mathbf{j} = e(n^+\mathbf{u}^+ - n^-\mathbf{u}^-)$,
the bulk velocity $\mathbf{u} = (n^+m^+\mathbf{u}^+ + n^-m^-\mathbf{u}^-)/\mu$,

$\mathbf{u}$ being defined so that $\mu\mathbf{u}$ is the momentum density.

With these definitions (2.6) yields immediately the laws of conservation of charge and mass, the latter being

$$\partial\mu/\partial t = -\operatorname{div}(\mu\mathbf{u}). \tag{2.14}$$

One of the equations resulting from (2.13) is obtained by adding the two equations and using (2.14). The result may be written

$$\mu\frac{\mathrm{d}u_i}{\mathrm{d}t} = -\frac{\partial S_{ij}}{\partial x_j} + (\rho\mathbf{E} + \mathbf{j}\wedge\mathbf{H}/c)_i, \tag{2.15}$$

where $$\mathrm{d}/\mathrm{d}t = \partial/\partial t + \mathbf{u}.\nabla,$$

and $$S_{ij} = T^+_{ij} + T^-_{ij} - \mu u_i u_j,$$

which reduces to

$$S_{ij} = S^+_{ij} + S^-_{ij} + \mu^{-1}n^+n^-m^+m^-(u^+_i - u^-_i)(u^+_j - u^-_j).$$

The last term represents a stress due to the relative motion between the protons and electrons, and it is important only when the current density is large. Equation (2.15) is the well-known equation of transfer of momentum and its validity does not depend on the frequency of collisions. Further the collision term involving $\delta' f$, which occurs in the equation for one constituent, but has been omitted in this section, does not contribute to (2.15), because the momentum lost by one constituent in collisions with the other is equal to the momentum gained by the other.

2.7. Ohm's law

The other equation resulting from (2.12) is obtained by multiplying by $\pm e/m^\pm$ before adding the two equations. After some rearrangement it becomes

$$\frac{m^+m^-}{e}\frac{\partial j_i}{\partial t} = e\mu(\mathbf{E} + \mathbf{u}\wedge\mathbf{H}/c)_i - (m^+ - m^-)(\rho\mathbf{E} + \mathbf{j}\wedge\mathbf{H}/c)_i + \frac{\partial}{\partial x_j}(m^+T^-_{ij} - m^-T^+_{ij}). \tag{2.16}$$

When the collision terms $\delta' f$ are included, (2.16) becomes the appropriate form of Ohm's law for totally ionized hydrogen, and completes the equations of motion. The left-hand side involving $\partial \mathbf{j}/\partial t$ is not usually included in Ohm's law and it was shown in § 1.3 that it can be neglected in cosmic electrodynamics.

The considerations of § 1.3 justify the neglect of the term involving $\rho\mathbf{E}$ in (2.16). The term involving $\mathbf{j}\wedge\mathbf{H}$ is the Hall electric field, which is perpendicular to both $\mathbf{H}$ and $\mathbf{j}$. The Hall effect in an ionized gas is analogous to that in a metal and results from the difference between the masses of electrons and protons, as is shown by the factor $(m^+ - m^-)$ occurring in this term. No important consequence of the Hall effect has yet been discovered in cosmic phenomena, though it is of major importance in the ionosphere. Since, from § 1.3, $n^+ \approx n^-$, the superscript on n can be omitted. The estimate of j/ne given in Table 1 shows that the Hall field is usually small compared with the dynamo field $-\mathbf{u}\wedge\mathbf{H}/c$, but it should not be entirely forgotten and will be further discussed in later chapters. The factor $(m^+ - m^-)/(m^+ + m^-)$ may be omitted, however, and then the Hall electric field $\approx \mathbf{j}\wedge\mathbf{H}/nec$.

The terms in (2.16) involving $T^{\pm}_{ij}$ represent the thermoelectric and similar effects. Rearranging them as in

$$m^+T^-_{ij} - m^-T^+_{ij} = m^+S^-_{ij} - m^-S^+_{ij} + m^+m^-(n^-u^-_iu^-_j - n^+u^+_iu^+_j)$$

it can be said that the first term on the right side is generally the largest, since it is $\sim m^+p$, while the terms involving $\mathbf{u}^{\pm}$ nearly cancel unless the current density is large. The resulting electric field is therefore $\sim kT/ea$, where k is Boltzmann's constant, and Table 2 contains numerical values of this field strength. It is seen that the

TABLE 2

	Sunspots	Corona	H II regions
a (cm.)	10^8	10^9	10^{11}
u (cm./sec.)	10^6	10^7	10^6
H (gauss)	10^3	10^2	10^{-5}
n (particles/c.c.)	10^{12}	10^7	1
T (° K.)	10^4	10^6	10^2
$300uH/c$ (volts/cm.)	10	10	10^{-7}
$300kT/ea$ (volts/cm.)	10^{-8}	10^{-7}	10^{-13}
σ/c^2	10^{-8}	10^{-5}	10^{-11}
$c^2/4\pi au\sigma$	10^{-7}	10^{-12}	10^{-7}
$\sigma uH/nec$ (cm./sec.)	10^9	10^{17}	10^{10}

thermoelectric field is much smaller than the induction electric field and the thermoelectric term will generally be omitted in this work. In situations where the mean free path is long and the velocity distributions not even approximately Maxwellian, the corresponding effect would still be weak, provided that no component of S_{ij} were very much larger than the appropriate value of nkT or equivalently, if the random velocities were not very much larger than the corresponding thermal velocities.

Equation (1.1) has now been obtained from (2.16), but the collision term has yet to be quoted from Cowling (1945) and shown to be small. The contribution to $\partial \mathbf{j}/\partial t$ of the collision term $\delta^1 f$ is $-\nu_{ep}\mathbf{j}$, where ν_{ep} is the effective frequency for collisions between electrons and protons. From (2.16) the conductivity σ is related to ν_{ep} by the equation

$$\sigma = e^2 n / m_e \nu_{ep}.$$

Cowling's value for σ in Gaussian units is $\sigma = c^2 T^{\frac{3}{2}}/K$, where

$$K = \frac{(8\pi)^{\frac{1}{2}}}{3} \frac{e^2 m_e^{\frac{1}{2}} c^2}{k^{\frac{3}{2}}} A_1(2),$$

and
$$A_1(2) \approx 2 \log \frac{4kT}{e^2 (2n)^{\frac{1}{3}}}$$

(see Chapman and Cowling (1939), p. 179 or, for some further discussion, the appendix).

For all our applications $A_1(2)$ is between 10 and 100, and it is sufficiently accurate for our purposes to take $K \sim 10^{14}$. The values of σ/c^2 given in Table 2 correspond to this value of K. The quantity $c^2/4\pi a u \sigma$ gives the order of magnitude of the ratio of the Ohmic electric field $\mathbf{j}/\sigma$ to the induction electric field $-\mathbf{u} \wedge \mathbf{H}/c$ and is seen to be very small, so justifying the neglect of the Ohmic field and hence the validity of equation (1.1). It may be noted that $\sigma u H/nec$ is the velocity difference between the electrons and protons required to make the Ohmic field comparable to the induction field; this is included in Table 2 and is seen to be extremely large (larger than the velocity of light in the case of the corona). Consequently, the assumption that the velocity difference between the electrons and protons is smaller than the thermal velocity of the electrons fails before the Ohmic field is comparable to the induction field. While the conductivity depends on n only through K, however, this velocity is inversely proportional to n and so is not so very large in the interior of a star (remember that the density given in Table 2 under sunspots refers to the chromosphere). Although it is so small, the Ohmic field may possibly be of importance either when the effective scale length is much smaller than usual or when a very long time is available for it to take effect. In § 3.4 the effect of all the terms in (2.16) will be further discussed using the concept of the motion of the magnetic field.

2.8. Viscosity

Viscosity may be regarded as friction occurring in the gas when its flow is non-uniform, but it is essentially an effect of diffusion. Consider a stream in which $\mathbf{u}$ has everywhere the same direction, but varies in magnitude across the stream. Then particles coming from one side of a given stream-line have on the average larger velocity components in the direction of $\mathbf{u}$ than particles coming from the other side, so that for particles arriving at this stream-line, the components of velocity parallel to $\mathbf{u}$ are correlated with the components perpendicular to $\mathbf{u}$. The distribution function is therefore anisotropic and the viscous force is obtained from the anisotropy of the stress terms in (2.9) and (2.15). The degree of anisotropy set up is controlled by its rate of reduction by collisions, as explained in § 2.5, and the coefficient of viscosity is therefore inversely proportional to the collision frequency. The kinematic viscosity (coefficient of viscosity/μ) will here be denoted by η, since its usual symbol ν is used for collision frequency. The viscous stress is given by

$$S_{ij}-p\delta_{ij}=-\mu\eta\left(\frac{\partial u_i}{\partial x_j}+\frac{\partial u_j}{\partial x_i}-\frac{2}{3}\frac{\partial u_k}{\partial x_k}\delta_{ij}\right) \tag{2.17}$$

and

$$\eta=kT/\alpha m\nu, \tag{2.18}$$

where α is the same as in § 2.5; the mean free path method gives $\alpha=\frac{1}{4}\pi$ and the Chapman–Enskog method gives $\frac{5}{8}$ for rigid spherical molecules. For the Coulomb interaction α depends slightly on the density and temperature, but for any of the values in Table 2 is very nearly equal to $\frac{1}{2}$ (Chapman and Cowling, 2nd ed., p. 179). The formula (2.18) refers to a simple gas with only one constituent, and requires modification for a mixture. For ionized hydrogen, however, corrections proportional to m_e/m_p may be neglected, and the viscous term in (2.15) is given quite accurately by (2.18), if $m_p \nu_{pp}$ is used for $m\nu$ (Chapman, 1954). There are also viscous terms in (2.16), and here the difference between the viscous *accelerations* of electrons and protons is involved. Since $\nu_{ee} = \surd 2\, \nu_{ep}$, the order of magnitude of η for the electrons is given by (2.18) with $m_e \nu_{ee}$ for $m\nu$, and this is $(m_p/m_e)^{\frac{1}{2}}$ times larger than η for protons. Because the viscous acceleration is greater for electrons than protons, a non-uniform flow of the gas causes an electromotive force in the same direction as the viscous force; this effect is considered further in §3.5.

So far the effect of the magnetic field on the viscosity has been ignored. All transport phenomena, such as viscosity and thermal conductivity, are affected by a magnetic field, and this is important when the relevant gyro-frequency is comparable with the relevant collision frequency. The effect on the interdiffusion of different constituents is illustrated by the Hall effect. The rate of transport of charged particles across the field is reduced by the magnetic field by a factor of approximately $\nu_{pp}/(\Omega_p^2 + \nu_{pp}^2)^{\frac{1}{2}}$, the rate of transport parallel to the field remaining unchanged. The viscosity is therefore reduced for flows in which the velocity varies in directions perpendicular to **H**, but not when it varies in the direction of **H**; thus in calculating the attenuation of Alfvén waves (see §5.4) travelling in the direction of **H**, the viscosity does not need modification.

2.9. Non-thermal conditions

It has already been stressed that the validity of the Boltzmann equation does not depend on the velocity distribution being approximately Maxwellian, and a few remarks will now be made about problems in which it is not; such conditions generally occur when collisions are infrequent, because collisions always make the distribution more nearly Maxwellian. Since the Boltzmann equation is still valid the formal development of this chapter is still correct, the only difference being that the stress tensor $-S_{ij}$ does not reduce to the form $-p\delta_{ij}$. It is possible to obtain an equation for dS_{ij}/dt by multiplying the Boltzmann equation by $v_i v_j$ before integrating over velocity space, but this equation involves the third moments of the velocity, that is $\int_{-\infty}^{\infty} v_i v_j v_k f \, \mathbf{dv}$, and so this procedure does not lead to a closed system of equations. No general method is known for solving such problems, and only specially simple problems have been solved. Nevertheless, it has been seen in § 2.7 that the stress terms in the Ohm's law equation (2.16) are generally small, in which case this equation is still useful.

The effect of a magnetic field on the stress tensor belonging to a charged constituent may be described by a rotation in velocity space with the gyrofrequency appropriate to that constituent. Since this frequency is usually large compared with the macroscopic time scale it may be expected that the velocity distribution is approximately symmetric about the direction of **H**; in this case one principal axis is parallel to **H** and S_{ij} may be described by just two quantities $S_{\parallel}$ the stress in the direction of **H** and $S_{\perp}$ the stress in any direction perpendicular to **H**. The time derivatives of these quantities are still difficult to determine, however. Because the mean free path is long, viscosity is very important, and for the same reason, it cannot be described by a coefficient of viscosity. In the case of the propagation of Alfvén waves of frequency larger than the collision frequency it is found that viscosity increases the effective inertia (Dungey, 1954).

The only problem involving non-thermal conditions to be discussed in the following chapters is the Chapman–Ferraro model for magnetic storms (§ 8.3). Here it is allowable to neglect the random velocities of the particles in the first approximation, so that there is no stress. This problem is in fact solved by calculating the orbits of the particles, all of which have the same initial velocity.

CHAPTER 3

THE MOTION OF MAGNETIC FIELDS

3.1. Representations of a magnetic field

The familiar method of representing a magnetic field by lines of force provides a picture that is particularly useful in cosmic electrodynamics. As so much use is made of this picture, it is perhaps necessary to remind the reader that the use of lines of force is a mathematical device and that they are not physical objects; the motion of lines of force is a further device, which is to be explained in this chapter, and it will be found that their motion is to some extent arbitrary, so that part of the motion can have no physical meaning.

The Maxwell equation, which has not yet been used, is

$$\operatorname{div} \mathbf{H} = 0. \tag{3.1}$$

This shows that the lines of force have no ends and makes possible the use of tubes of force, a similar mathematical device, developed by Faraday, but not much used in modern laboratory electrodynamics. A tube of force may be defined as the surface generated by the lines of force passing through a closed curve; the simplest picture is obtained by choosing a small closed curve, so that a narrow tube is obtained. From the definition, the component of the field normal to the surface of a tube of force vanishes. It then follows from (3.1) that the magnetic flux $\int \mathbf{H}.dS$ through any cross-section of a given tube of force is a constant, which may be called the flux of the tube. Thus, whereas a line of force gives only the direction of the field, a narrow enough tube of force gives complete information about the field strength as well, the strength in different parts of the same tube being inversely proportional to the area of cross-section. By picturing narrow tubes of force as elastic strings a useful picture of the effect of the magnetic force-density is also obtained, but this belongs to the next chapter. It may be noted that, when the field is pictured in terms of lines of force, it is not practicable at the same time to describe it as the superposition of several fields with different sources, as is done in many branches of electrodynamics; as we shall be using lines of force, this type of analysis will not be used, and the only field mentioned here will be the total field.

Equation (3.1) allows the representation of the magnetic field by a vector potential **A** such that

$$\mathbf{H} = \operatorname{curl} \mathbf{A}. \tag{3.2}$$

The vector potential is not very useful in cosmic electrodynamics, except in discussing specially simple models of the two-dimensional type, in which **A** and also **j** have only one non-zero component each. Another representation, which is more useful for general problems and equivalent to the vector potential for two-dimensional problems is

$$\mathbf{H} = F \nabla\phi \wedge \nabla\psi \equiv F \operatorname{curl}(\phi\nabla\psi), \tag{3.3}$$

where ϕ and ψ are independent functions of position, each of which is constant along any line of force, and F is a function of position which is required by (3.1) also to be constant along a line of force and therefore to be a function of ϕ and ψ only. It follows that the lines of force are the curves in which the surfaces of constant ϕ intersect the surfaces of constant ψ, and that the tube of force bounded by the four surfaces corresponding to the particular values ϕ, $\phi+\delta\phi$, ψ, $\psi+\delta\psi$ contains an amount of flux given by $F\delta\phi\delta\psi$.

There is a type of magnetic field for which two independent families of surfaces, each containing lines of force, do not exist and for these (3.3) needs modification; it is best explained by an example. The term 'linking' will be applied to lines of force and has its everyday meaning. To be precise let a closed curve C enclose a surface S; then if a line of force crosses S in one direction N_1 times and in the other N_2 times, the line of force is linked $(N_1 - N_2)$ times with C and is unlinked with C only if $N_1 = N_2$. Consider a twisted toroidal field, in which each line of force lies on a toroidal surface and is linked with the lines of force inside this surface. In general each line of force is infinite in length and covers the whole of the toroidal surface; this is possible because the surface is not simply connected, so that the line need not cross itself. Then the only surfaces containing lines of force are the toroidal surfaces; any other surface generated by lines of force fills up a finite volume.

The representation (3.3) can still be used, however, if a cut is made, such that the surface covered by any line of force is made to be singly connected; on a toroidal surface the cut may be taken to be a closed curve linking the toroid. Then, supposing that ϕ is constant on the surface, ψ may vary, but will be discontinuous at the cut. The field is given by (3.3) except on the surface of the cut. The cut has no physical significance and the field is continuous and is given by (3.3) on either side of the cut; thus (3.3) can be used without any practical difficulty. It will be seen later that this type of field, in which a line of force covers a whole surface, is the general type of actually existing fields. The representation (3.3) will be used, with the assumption that suitable cuts are made.

3.2. The motion of the field with the material

In this chapter we consider the temporal variation of the magnetic field. This is only a part of the general cosmic electrodynamic problem, but it can be treated separately. We recall from our causal scheme that $\partial\mathbf{H}/\partial t$ is determined by equation (1.5) and $\mathbf{E}$ by (1.4) (or (2.16)). We now come to the most useful general result in cosmic electrodynamic theory. It was seen in §2.7 that in the majority of astrophysical applications the contribution to $\mathbf{E}$ from the induced field $-\mathbf{u}\wedge\mathbf{H}/c$ is very much greater than that from any of the other terms, so that (1.1) is a good approximation; as already noted this corresponds in terms of circuits to the dominance of inductance. In this section we suppose that (1.1) is exactly true so that from (1.5)

$$\frac{\partial\mathbf{H}}{\partial t}=\operatorname{curl}(\mathbf{u}\wedge\mathbf{H}) \tag{3.4}$$

and show that the magnetic field is then 'frozen into the material'. The effect of the other terms in (2.16) on this result is discussed later in this chapter.

The simplest proof of the freezing-in of the field (Sweet, 1950) makes use of the representation (3.3). Suppose that the surfaces of constant ϕ and of constant ψ move with the velocity $\mathbf{u}$ and that F remains always the same function of ϕ and ψ. These statements are expressed by the vanishing of $d\phi/dt$, $d\psi/dt$ and dF/dt, where $d/dt=\partial/\partial t+\mathbf{u}.\nabla$. Let us then replace $\mathbf{u}.\nabla$ by $-\partial/\partial t$ in the identity

$$\mathbf{u}\wedge\mathbf{H}=F((\mathbf{u}.\nabla\psi)\nabla\phi-(\mathbf{u}.\nabla\phi)\nabla\psi)$$

and operate with curl, giving

$$\operatorname{curl}(\mathbf{u}\wedge\mathbf{H}) = F(\nabla\phi\wedge\nabla\,\partial\psi/\partial t+\nabla\,\partial\phi/\partial t\wedge\nabla\psi)-F^{-1}(\mathbf{u}\wedge\mathbf{H})\wedge\nabla F.$$

Now $$F^{-1}(\mathbf{u}\wedge\mathbf{H})\wedge\nabla F = F^{-1}(\mathbf{u}\,.\,\nabla F)\,\mathbf{H} = -\frac{\partial F}{\partial t}(\nabla\phi\wedge\nabla\psi),$$

using $\mathbf{H}\,.\,\nabla F = 0$ in the first step.

Therefore

$$\operatorname{curl}(\mathbf{u}\wedge\mathbf{H}) = \frac{\partial}{\partial t}(F\,\nabla\phi\wedge\nabla\psi).$$

Consequently, when (3.4) is true, the representation (3.3) is maintained for all time, if the temporal change of ϕ, ψ and F is described by the motion $\mathbf{u}$ in the way supposed. In this picture the lines and tubes of force, which are determined by the surfaces of constant ϕ and ψ, are to be visualized as moving with the velocity $\mathbf{u}$; they can be labelled with their values of ϕ and ψ and can be followed through the motion. A tube of force identified in this way contains a constant amount of magnetic flux $F\delta\phi\,\delta\psi$; a segment of a tube of force experiences the same strain as the material it contains and consequently the field strength varies with the time inversely as the area of cross-section. This may alternatively be seen by writing (3.4)

$$\mathrm{d}\mathbf{H}/\mathrm{d}t = (\mathbf{H}\,.\,\nabla)\,\mathbf{u} - \mathbf{H}\operatorname{div}\mathbf{u}. \qquad (3.4')$$

The first term on the right-hand side shows that the field strength is increased by stretching the tube without changing the density of the material. The second term is proportional to the rate of change of density and the combination of these two terms is proportional to the rate of change of cross-section. The first term also rotates the direction of the field when the material is rotating.

The result obtained in this section is frequently referred to as the 'freezing' of the magnetic field into the material, and it is exactly equivalent to (3.4). By picturing lines of force in motion, the following consequences of (3.4) are easily seen:

(i) the quantity of material in any tube of force remains constant in time;

(ii) no pair of lines of force can become linked or unlinked in the course of the motion;

(iii) since the flux in each tube is constant in time, flux cannot be created or destroyed. (A tube cannot be shrunk to a point because of (i).)

When (1.1) is a good approximation it is likely that the changes forbidden by these conclusions can only proceed slowly, as the field only slightly 'slips' relative to the material. Slow 'slipping' may, however, be important in problems of stellar evolution, and special circumstances, for which (1.1) is not even approximately true, may also be important. As the picture of the field in motion is so useful in this subject, we next consider how the temporal change of the magnetic field can be represented by a motion, when (1.1) is not true; the departure of the electric field from that given by (1.1) is then represented as a motion relative to the material.

3.3. Motion of the field in the general case

It has been seen in the last section that the statement that a magnetic field moves with velocity $\mathbf{w}$ means

$$-c\,\mathrm{curl}\,\mathbf{E} \equiv \partial\mathbf{H}/\partial t = \mathrm{curl}\,(\mathbf{w}\wedge\mathbf{H}), \tag{3.5}$$

but, if the electric field $\mathbf{E}$ is given, $\mathbf{w}$ is not completely determined by (3.5); any velocity $\mathbf{w}'$ such that $(\mathbf{w}'-\mathbf{w})\wedge\mathbf{H}$ is irrotational would also satisfy (3.5) and could be used as the velocity of the field just as well as $\mathbf{w}$. (When (1.1) *is* true it is obviously most convenient to take the velocity of the field to be the velocity $\mathbf{u}$ of the material). The component of $\mathbf{w}$ parallel to $\mathbf{H}$ is arbitrary and therefore has no physical significance. The component of $\mathbf{w}'-\mathbf{w}$ perpendicular to $\mathbf{H}$ must have the form $H^{-2}\mathbf{H}\wedge\nabla W$, where W is constant on any line of force, but is otherwise arbitrary. Such a velocity $\mathbf{w}'-\mathbf{w}$ does not change $\mathbf{H}$, but just permutes the lines of force.

The temporal change of a magnetic field cannot always be described as a motion of the field. Suppose that there are lines of force not covering a whole surface; integrating (3.5) over a surface bounded by a line of force and using Stokes's theorem

$$-c\oint \mathbf{E}\,.\,\mathbf{ds} \equiv \int \partial\mathbf{H}/\partial t\,.\,\mathbf{dS} = \oint (\mathbf{w}\wedge\mathbf{H})\,.\,\mathbf{ds}. \tag{3.6}$$

For the surface bounded by a line of force $\int \partial\mathbf{H}/\partial t\,.\,\mathbf{dS} = \mathrm{d}/\mathrm{d}t \int \mathbf{H}\,.\,\mathbf{dS}$ and the right-hand side of (3.6) vanishes; hence the magnetic flux linked by a line of force must be constant; this was conclusion (ii) of § 3.2, obtained by picturing a field in motion. The left-hand side of (3.6), however, does not in general vanish, and therefore the concept of the motion of the field is not always valid.

This difficulty can be overcome by choosing a surface S which is crossed at least once by each line of force (the area of S may be infinite) and by regarding the lines of force as being 'disconnected' on S. The lines can be pictured as being cut at this surface, the two ends arising from the same cut being allowed to move on S with different velocities. Their motion must be such that $\mathbf{H}$ remains continuous and hence their relative motion must just permute the ends of the lines of force; at any boundary of S their relative motion must be parallel to the boundary. S may consist of a set of separate surfaces whose total area may be infinite and in general these must be given a motion such that S continues to intersect every line of force. This picture is of course just a mathematical device and S has no physical significance.

The motion of the field can now be defined as follows. Let

$$V = \int \mathbf{E} . d\mathbf{s}, \tag{3.7}$$

the integral being taken along a line of force in the direction of $\mathbf{H}$, starting from S and not crossing S again. Then, everywhere except on S, $\mathbf{E} - \nabla V$ is perpendicular to $\mathbf{H}$ and (3.5) is satisfied with

$$\mathbf{w} = cH^{-2}(\mathbf{E} - \nabla V) \wedge \mathbf{H}. \tag{3.8}$$

At S both V and $\mathbf{w}$ may be discontinuous, but on both sides of S, $\partial\mathbf{H}/\partial t$ is equal to $-c \operatorname{curl} \mathbf{E}$, which is continuous.

If, as in the case of the twisted toroidal field, some lines of force cover a whole surface, they will cross the surface S an infinite number of times. A line of force which crosses S more than once is cut into sections, which could be labelled and followed through the motion. The connexions at S between the sections are altered by the motion, so that two sections, which at one instant form part of the same line of force, may belong to different lines of force at another instant. Similarly, a line of force, which at one instant crosses S only once, may at another instant be just a section of a line of force which crosses S more than once. Thus the lines of force do not generally retain their identity during the motion, though the sections do. It now follows that the type of field in which each line of force covers a whole surface is the general type. Given a field for which the lines of force are finite in length, the slightest reconnexion at the surface S removes this property, and the lines of force remain finite in length only in the singular cases, for which the ends happen to join up.

3.4. Causes of relative motion between the field and the material

Some of the causes of relative motion between the field and the material are represented by those terms in (2.16) which are neglected in (1.1). The term which usually contributes the largest velocity is the Hall electric field. This is approximately $\mathbf{j} \wedge \mathbf{H}/nec$ and so is always perpendicular to $\mathbf{H}$ and its effect is to add a velocity $\mathbf{j}/ne$ to $\mathbf{w}$ with no contribution to the discontinuity at S. Consequently the conclusion (ii) of § 3.2 is not affected by the Hall effect; conclusion (i) is hardly affected because the change of mass due to the Hall field is proportional to the change in the total charge contained by the tube of force. Using the exact form of the Hall field given in (2.16) the nett velocity of the field is given by

$$\mathbf{w} = \frac{m^+\mathbf{u}^- + m^-\mathbf{u}^+}{m^+ + m^-},$$

which may be described by saying that the field moves more closely with the electrons than with the protons, their velocities being weighted inversely with their masses.

It was seen in § 2.7 that the Ohmic field $\mathbf{j}/\sigma$ is usually very small, but its effect may occasionally be of interest. Its representation in terms of the motion of the field is rather complicated, but (1.3) and (1.5) show that its contribution to $\partial\mathbf{H}/\partial t$ is

$$-(4\pi)^{-1} c^2 \operatorname{curl}(\sigma^{-1} \operatorname{curl} \mathbf{H}).$$

The effect of this term may be described as a 'diffusion' of the field, and it has the effect of smoothing out small irregularities in $\mathbf{H}$ with a time constant of $c^2/4\pi\sigma b^2$, where b is their scale length. The effect of spatial variation of the conductivity in the above expression tends to concentrate the current in the regions of high conductivity, but this effect is also slow, if the scale is large and the conductivity is large everywhere. It may also be noted that the Ohmic field determines the rate of Joule heating, the field energy being reduced at the same rate. The Ohmic field can contribute to V and so can alter the linking of the lines of force and destroy flux (see § 3.6).

The terms involving $T_{ij}^{\pm}$ in (2.16) can contribute to V, but have been seen to be small. It may further be pointed out that under thermal conditions the largest term is a constant multiple of ∇p; if, in addition, $\nabla n \wedge \nabla T = 0$, this term is irrotational and does not contribute to the motion of the field.

In (2.16) complete ionization is assumed and we next describe three additional effects which result from the presence of neutral particles. First, the resistivity is increased by collisions between charged and neutral particles; this is seldom important, however, because even with only 0·1 % ionization collisions between charged particles are the more frequent for temperatures less than 10^4 ° K, while at higher temperatures the ionization is nearly complete.

A second effect occurs if the mean velocity of the neutral particles differs from $\mathbf{u}$; although such a velocity difference is opposed by collisions between the neutral and charged particles, it may be appreciable. Because of their smaller mass, the electrons are affected more than protons by collisions with neutral particles, so that there is a contribution to $\partial \mathbf{j}/\partial t$ in the direction opposite to the relative velocity of the neutral particles; the effect is small and is discussed further by Hoyle (1949, appendix II).

Another effect results from ionization and recombination; the Boltzmann equations, being conservation equations, require additional terms to allow for these. A relative flow between the material and the field can result if ions recombine, move across the field, and are then reionized. When this occurs, conclusion (i) of § 3.2 is obviously invalidated, (ii) remains valid, and (iii) is invalidated in so far as it depends on (i). This effect may be expected to be important in cosmic electrodynamic phenomena, as there is often sufficient time for recombination and ionization to occur.

This section is concluded with a list of causes of relative motion between the field and the material, in which are included, for the sake of completeness, the effect of radiation pressure, which is probably unimportant, and the enhancement of the Ohmic decay by turbulence, which is to be explained in § 5.8.

List of causes of relative motion between the field and the material

Hall effect.
Electrical resistance.
Enhancement of resistance by turbulence.
Thermoelectric effects.
Viscous effect (see § 2.8).
Radiation pressure.
Relative motion of neutral particles.
Recombination and ionization.

3.5. The generation of a magnetic field by mechanisms involving non-uniform rotation

As a rotating star condenses, there are several mechanisms, which cause electric fields perpendicular to the meridian planes, and hence the growth of magnetic fields, which qualitatively resemble the dipole form. One such effect is that of viscosity, mentioned in § 2.8. Another involves the radiation pressure; if the angular velocity decreases outwards, the radiation coming from inside and arriving at distance r from the centre corresponds to an angular velocity greater than that at r and consequently the radiation pressure tends to increase the angular velocity. As in the case of viscosity, the acceleration due to radiation pressure is not the same for electrons and protons, and the difference must be balanced by an electric field. There may be other similar mechanisms, but we here obtain an upper limit to the magnitude of the field generated, which is independent of the mechanism, provided that the directions in which the electrons and protons are accelerated are the same. Consider an axially symmetric model and let electrons and protons be subject to forces F_e and F_p perpendicular to the meridian plane. The electric field E_1 needed to make their accelerations equal is given by

$$E_1 = \frac{m_p F_e - m_e F_p}{e(m_p + m_e)}$$

and, since F_e and F_p have the same sign,

$$|E_1| < \frac{m_p |F_e + F_p|}{e(m_e + m_p)}.$$

Then integrating round a circle of symmetry and using (1.5) gives

$$\left|\frac{\partial_1}{\partial t}\int \mathbf{H}.d\mathbf{S}\right| < \frac{m_p c}{e}\left|\frac{\partial_1}{\partial t}\oint \mathbf{u}.d\mathbf{s}\right|,$$

where $\partial_1/\partial t$ means the contribution of F_e and F_p to $\partial/\partial t$. Now let the circle of the integration contract with the velocity of condensation of the star; the results of § 3.2 show that, if **E** were just the induced field, $\int \mathbf{H}.d\mathbf{S}$ would be constant. Similarly, Thomson's theorem states that $\oint \mathbf{u}.d\mathbf{s}$ is constant, when there is no force. Consequently an upper limit for the magnetic flux generated is given by $m_p c/e \times$ the change in circulation round a circle which condenses with the star.

If no condensation occurred the average magnetic field generated in the circle would be no greater than $2m_p c/e \times$ the change in angular velocity and $2m_p c/e = 2.10^{-4}$ gauss sec. The cosmically occurring values of angular velocity therefore correspond to very small field strengths. When a star condenses, however, much stronger fields may result. It is believed that most of the original angular momentum is lost, and that the circulation round a circle condensing with the star is reduced by many orders of magnitude. The field generated may then be of the order $2m_p c/e \times$ the angular velocity which would have existed, if no angular momentum had been lost. Even so extreme values must be supposed for the cloud out of which a star condensed, if a field strength as large as 0·1 gauss is to be obtained.

3.6. The dynamo problem

A question that has received considerable attention concerns the possibility of a mechanism, analogous to the self-excited dynamo, occurring in cosmic systems; can the electric field induced by motion in an existing magnetic field maintain this field against Ohmic loss? This problem has received particular attention in connexion with the geomagnetic field (Elsasser, 1956*b*), but only the general problem is considered here. Cowling (1934) used a formulation very similar to that of § 3.3 in discussing sun-spots; he simplified the problem by imposing axial symmetry. With symmetry of this kind, simple results follow directly in terms of the motion of the field. Suppose that the lines of force all lie on planes (in the case of axial symmetry the meridian planes), so that no line of force can cover a whole surface and each line of force is just a closed loop. Suppose further that the symmetry is maintained; then the motion of the field can be chosen such that each line of force stays on its plane. No lines of force are linked, the lines or tubes of force can be followed through the motion and only one plane need be considered, so that the problem is two-dimensional. It is easily seen, by picturing two-dimensional motion, that lines of force cannot appear or disappear except at a neutral point, where $\mathbf{H}=0$. Two types of neutral point will be distinguished: the lines of force in the neighbourhood of a neutral point may be elliptical or hyperbolic, and the neutral point will be called O-type in the former case and X-type in the latter case. At an 'O-type' neutral point lines of force can grow out of the neutral point or shrink into it. Since $w \propto H^{-1}$, flux is created or destroyed at a finite rate unless $\mathbf{E}$ vanishes at the neutral point; in three dimensions there is a neutral line corresponding to the neutral points in the planes of symmetry and the rate of change of magnetic flux from a section of the neutral line is equal to $c\int \mathbf{E}.\mathbf{ds}$ taken along it, which of course follows from a well-known theorem. At an 'X-type' neutral point, flux is created by the breaking of a line of force to form two lines or destroyed by the reverse process; this mechanism will be described in more detail in §6.4. Since, for fields restricted to this type of symmetry, flux is created and destroyed only at neutral lines, it should be possible to define the 'total flux' of the field; this can be done by dividing the field into sufficiently small tubes of force and summing the magnitudes of the flux in each.

The effect of the several terms in (2.16) at neutral points may be briefly considered. The Hall field has no effect, since it is proportional to H. The diffusion effects can either create or destroy flux. At an O-type neutral point, however, curl $\mathbf{H}$ always has the direction such that the Ohmic field destroys flux. At an X-type neutral point the Ohmic field may create or destroy flux. Further, any field with the type of symmetry assumed above must have at least one O-type neutral point; this follows from the representation in which the lines of force are contours of constant ϕ; since these are closed, ϕ must have a maximum or minimum, which corresponds to an O-type neutral point. The self-excited dynamo, requiring a steady state in which the Ohmic field is balanced by the induction field, is therefore impossible, because the Ohmic field cannot be so balanced at an O-type neutral point.

In attempting to generalize these considerations to fields with no special symmetry, so that a line of force may cover a whole surface, some difficulty is met. No precise definition of the total flux will be given, but in the case of a twisted toroidal field, the flux between neighbouring toroidal surfaces is obviously a significant quantity and for many types of field the total flux is made up of such parts. It may be noted that the flux through an arbitrary surface such as a sunspot can be much larger than the 'total flux', if an individual tube of force crosses the surface many times.

With a field of the general type, flux can still be created and destroyed at neutral lines, but now there need not be any neutral lines. There is at least one other way in which flux can be created which is as follows. Suppose that two parts of the surface of a tube of force come into contact over a finite area; in practice the field will be continuous at the interface. Now choose S of §3.3 to be a surface crossing both parts of the tube where they are in contact. It is evidently possible in principle for the ends of the lines of force to move, so that the ends which originally belonged to one part of the tube become connected to the ends which originally belonged to the other part; if this occurs, two separate tubes are obtained, each containing the same amount of flux as the original tube. An attractive mechanism in which this occurs was suggested by Alfvén (1950*b*) and will be described in §4.5.

This example shows that care is sometimes needed in applying the picture developed in §3.3. If S had been taken to cross only one part of the tube, the correct result would not have been obtained. The reason is that the relative velocity of the ends of the lines of force at the boundary of S would not have been parallel to the boundary. Care is also needed in discussing any model containing finite regions of zero field, since magnetic flux can cross such a region at a finite rate.

3.7. Instability of unlinked lines of force

Equilibrium configurations of magnetic fields are discussed in the next chapter, but the ideas developed in this chapter yield the simple result that any field with no linked lines of force is unstable. We need only know that the field energy $\int \frac{H^2 \, dV}{8\pi}$ contained in a thin tube of force is proportional to its length and inversely proportional to its mean area of cross-section. It follows that the energy of a thin toroidal tube of force can be reduced by radial contraction accompanied by expansion in the direction of the axis as shown in fig. 3.1 (*a*); the tube becomes a long thin cylinder as in fig. 3.1 (*b*),

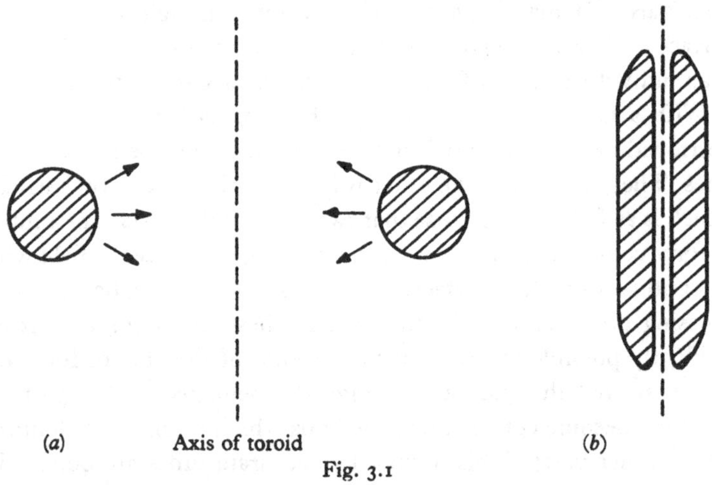

Fig. 3.1

and its volume need not have changed. If now there are no linked lines of force the tube can be split into smaller tubes, which can be separated and contracted, and the process can be repeated indefinitely. Thus the magnetic energy can always be reduced by motion, without compressing the material, and hence such a field is unstable.

It has been noted already that, in general, there are linked lines of force and the argument just given then fails. On the other hand, the simplest mathematical models, some of which are to be discussed in the next chapter, have no linked lines of force, and it should be remembered that some modification is needed to give these models stability.

CHAPTER 4
MAGNETOSTATICS

4.1. The magnetic force density

In this chapter we consider models in which the gas is static and in mechanical equilibrium. The equilibrium condition involves the magnetic field by way of the magnetic force density $\mathbf{j} \wedge \mathbf{H}/c$. Partly because this force density is non-linear in $\mathbf{H}$, most of the discussion of dynamic models can only be qualitative, and it is useful to have some quantitative illustrations of its effect; such illustrations are provided by static models. The variation of the magnetic field with time is ignored; it has been discussed in the last chapter, where it was seen that $\partial\mathbf{H}/\partial t$ is nearly always small when the gas is static, and the other half of the problem is now discussed separately, the two halves being brought together in the next chapter.

A further reason for investigating static equilibrium is that certain features observed on the sun do appear to approximate to static conditions. The gravitational field is known and observations provide some idea of the pressure gradient and of the acceleration of the gas. Then when the resultant of the pressure gradient and the force density due to gravitation is found to be much larger than the force density required by the acceleration (for instance if the acceleration is known to be much less than g and the pressure gradient is nearly horizontal), it may be inferred that this resultant is approximately balanced by the magnetic force density, in which case the condition of static equilibrium is approximately satisfied. In some cases this argument has been the main evidence for the existence and nature of the magnetic field, but direct evidence has recently been obtained by Babcock with his magnetograph.

There are several useful ways of describing the magnetic force density $\mathbf{j} \wedge \mathbf{H}/c$ (it was seen in chapter 1 that the force density $\rho\mathbf{E}$ due to space charge is negligible). We may consider the magnetic energy, whose density is $H^2/8\pi$. Chandrasekhar and Fermi (1953) have shown that the virial theorem is still true for a perfectly conducting fluid in a magnetic field, provided the magnetic energy is included. In a uniform expansion the total magnetic energy varies inversely as the scale, as does the gravitational energy; a magnetic star would therefore be unstable with respect to expansion, if the sum of the magnetic and gravitational energies were positive.

The magnetic energy in a thin tube of force containing the flux F is $F^2\oint A^{-1}\mathrm{d}l/8\pi$, where A is the area of cross-section, $\mathrm{d}l$ is the element of length and the integral is taken along the tube. When the volume and length are both fixed, the energy has its minimum value for the configuration in which A is constant along the tube, so that, if a narrow neck occurs in the tube, the energy is increased. When the volume is fixed and A is the same all along the tube, the energy is proportional to the square of the length. A similar representation is provided by the Maxwell stress tensor which is $(2H_iH_j-H^2\delta_{ij})/8\pi$ omitting the terms involving the electric field, since they are smaller by a factor $(u/c)^2$. The stress then consists of a tension of amount $H^2/8\pi$ parallel to the field and a lateral pressure of the same amount perpendicular to the field; alternatively, it may be regarded as an isotropic pressure $H^2/8\pi$ and a tension $H^2/4\pi$ parallel to the field. This description shows that a thin tube of force may be pictured as behaving like an elastic string, whose natural length is zero. It tries to reduce its energy by motion of the kind described in § 3.7. The elastic string picture is particularly useful in qualitative investigations; it shows that ionized gas in a magnetic field behaves rather like fibrous material, consisting of endless unbreakable fibres. Consider next an example of its 'fibrous behaviour'.

Suppose that a region of zero field surrounds a region containing a field, the surrounding material being at a pressure sufficient to keep the field compressed. If a disturbance makes a dent in the boundary of the two regions the lines of force will be distorted as in fig. 4.1 (*a*). The magnetic force will oppose the distortion, but remembering to think in three dimensions we see that the material in the field can easily split in a surface containing lines of force. Fig. 4.1 (*b*) shows the disturbance in a plane perpendicular to the field, the shaded part being the region containing the field. The lines of force at P can straighten themselves by moving round to Q. The energy is then not much greater than before the disturbance.

In this way a lump of external matter may penetrate into the region of the field and make a hole in the field. This hole will be squeezed by the field in directions perpendicular to the lines of force, but not in the direction parallel to them. It follows that the hole will spread in the direction parallel to the field, so that it takes the shape of a filament.

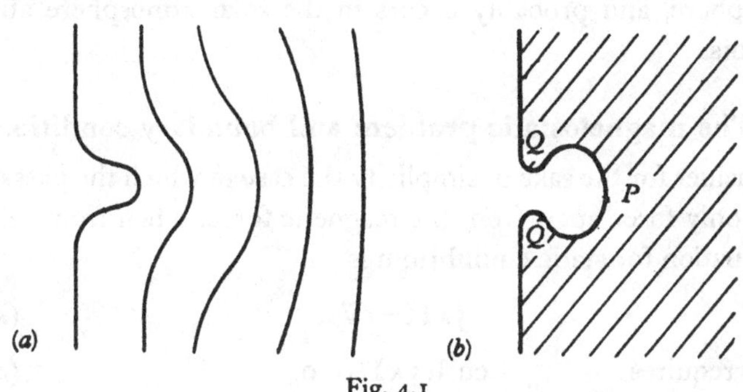

Fig. 4.1

Next let us consider the expression $\mathbf{j}\wedge\mathbf{H}/c$; this is obviously perpendicular to $\mathbf{H}$, a property which must be remembered as it will now be rewritten in a form for which it is not obvious. Using (1.3)

$$4\pi\mathbf{j}\wedge\mathbf{H}/c=(\mathbf{H}.\nabla)\mathbf{H}-\tfrac{1}{2}\nabla H^2, \tag{4.1}$$

which can alternatively be obtained from the Maxwell stress tensor. It is convenient to describe the second term as the gradient of the 'magnetic pressure' $H^2/8\pi$ and to define the 'total pressure' to be $\Phi/4\pi$, where

$$\Phi=4\pi p+\tfrac{1}{2}H^2. \tag{4.2}$$

The component of $(\mathbf{H}.\nabla)\mathbf{H}$ parallel to the field simply cancels the gradient of the magnetic pressure in this direction, since $\mathbf{j}\wedge\mathbf{H}$ is perpendicular to the field. The component of $(\mathbf{H}.\nabla)\mathbf{H}$ perpendicular to the field can be described geometrically as follows. At any point let $\mathbf{k}$ be a vector whose magnitude is equal to the curvature of the line of force and whose direction lies in the osculating plane of the line of force, is normal to the line of force and is directed towards the concave side. Then the component of $(\mathbf{H}.\nabla)\mathbf{H}$ perpendicular to $\mathbf{H}$ is $H^2\mathbf{k}$. In terms of the force density it is this part which tends to make a line of force contract.

It may be noted here, that, if the magnetic force density is balanced by the pressure gradient, the order of magnitude of the pressure variation is generally $H^2/8\pi$; if the pressure variation is substantial, then $H^2 \sim 8\pi p$. Conversely, if $H^2 \gg 8\pi p$ and the gas is in mechanical equilibrium, the component of $\mathbf{j}$ perpendicular to $\mathbf{H}$ must be smaller than that given by the usual order of magnitude of curl $\mathbf{H}$; this condition is known to occur in the earth's upper atmosphere, and probably occurs in the solar atmosphere above sunspots.

4.2. The magnetostatic problem and boundary conditions

Consider for the sake of simplicity the case in which the pressure is the only force apart from the magnetic force. Then from (2.15) the equation for static equilibrium is

$$\mathbf{j} \wedge \mathbf{H} = c\nabla p, \tag{4.3}$$

which requires

$$\operatorname{curl}(\mathbf{j} \wedge \mathbf{H}) = 0. \tag{4.4}$$

When (4.4) has been solved for $\mathbf{H}$, (4.3) determines p apart from an additive constant, which is restricted only by the condition that p is everywhere positive. Note also from (4.3) that p is constant on any line of force. In this problem the temperature is not involved; when there is a gravitational field the problem is more complicated and in addition some assumption must be made about the temperature.

Using (4.1), (4.4) becomes

$$\operatorname{curl}((\mathbf{H}.\nabla)\mathbf{H}) = 0. \tag{4.5}$$

This has the same mathematical form as the hydrodynamical equation

$$\operatorname{curl}((\mathbf{u}.\nabla)\mathbf{u}) = 0$$

for steady flow in an inviscid incompressible fluid of uniform density, but the solutions of (4.5) appear to be less useful in hydrodynamics than in magnetostatics, and have not been studied systematically.

There are of course a great variety of solutions of (4.5) and it is necessary to make some further restriction in order to solve it. Later we shall assume special symmetries, but first discuss another assumption which leads to a class of solutions which are useful because they can satisfy reasonable boundary conditions. Suppose that we could find a set of lines of force such that any magnetic field with these lines of force is a solution of (4.5); then the quantity of flux in each tube of force would be quite arbitrary and we could say that such a set of lines of force is specially favoured. Let **H** be a particular field whose lines of force have this property; and let $\mathbf{H}'$ be defined as $\lambda\mathbf{H}$ where λ is a scalar function of position, which is arbitrary except for the condition $\mathbf{H}.\nabla\lambda = 0$; then $\mathbf{H}'$ is another field with the same lines of force and (4.5) requires

$$\operatorname{curl}((\mathbf{H}'.\nabla)\mathbf{H}') = 0.$$

Now (4.3) can be written, using (4.2),

$$(\mathbf{H}.\nabla)\mathbf{H} = \nabla\Phi,$$

and then
$$(\mathbf{H}'.\nabla)\mathbf{H}' = \lambda^2\nabla\Phi,$$

and so we require
$$\nabla\lambda \wedge \nabla\Phi = 0.$$

This means that the surfaces of constant λ and of constant Φ are the same, or that λ is a function of Φ. This restriction on λ shows that the lines of force cannot have such a general property as we were seeking, except when $(\mathbf{H}.\nabla)\mathbf{H}$ vanishes, in which case the lines of force are straight lines. Sets of lines of force such that $\mathbf{H}'$ satisfies (4.5) when λ is an arbitrary function of Φ are still favoured lines of force, however; this property will be referred to as (A) for brevity. Since λ is constant on a line of force, so must Φ be, and then, since p is, H^2 is constant on a line of force; this is the condition on the field required for the property (A). The boundary conditions of magnetostatics now require some consideration, and it will be seen that solutions of (4.5) with the property (A) can be made to satisfy a physically acceptable boundary condition.

In this general kind of work there are no specific boundary conditions to be applied, and, if there were, the solution of (4.5) would be an awkward mathematical problem. Nevertheless, it is desirable to impose some conditions of reasonable behaviour. Interest may be confined to a finite region of space, but it would be disturbing if the magnetostatic equation led to singularities outside this region. The safest way to prevent this from occurring would be to impose the condition that H should vanish outside some region. The simplest model for which this can be done is the twisted toroidal field, for which the mathematical solution is intractable.

For a solution with the property (A), however, H can be made to vanish outside a surface of constant Φ by choosing $\lambda(\Phi)$ to vanish there, and, although the region enclosed by this surface is not finite, this condition is better than none. In the simplest example (§4.3(i)) the surfaces of constant Φ are infinite cylinders, but the field does not vary along the cylinders, and the model is not too unreal. The following discussion is not restricted to models with the property (A) however.

It is convenient for simplicity to consider models in which H changes discontinuously to zero at some surface of constant Φ. Such a discontinuity is of course physically unreal; it implies that there is a finite current flowing in the surface of discontinuity, and this current will spread out owing to the Ohmic electric field. We should, therefore, consider only models in which H never varies by a large factor in a distance less than b, where $4\pi\sigma b^2/c^2$ is a long time on the scale of some phenomenon, which the model is supposed to represent. Then the discontinuous model would have to be modified by adding a layer of width $\geqslant b$ outside the discontinuity, and making H vary smoothly in this layer, so as to reach the value zero outside it. In practice table 2 shows that b is much less than the characteristic length a of the model, so that this extra layer may still be thin; then the mechanical behaviour of this more correct model is practically the same as that of the discontinuous model, and, in discussing only the mechanical half of the problem, this extra layer may be omitted. It will be seen later that observations suggest that rapid spatial variations in $\mathbf{H}$ do sometimes occur.

4.3. Two-dimensional models

The simplest models are obtained when the lines of force lie in parallel planes and are the same in every plane. They are unstable in the way described in §3.7. Their symmetry is expressed in terms of the representation (3.3) by taking $\psi = z$, $\phi =$ a function of x and y and $F =$ a constant, so that the lines of force lie in the planes of constant z and are independent of z. Equivalently we may use a vector potential $(0, 0, A)$; A may then be equal to ϕ in (3.3) and is constant on a line of force. Using (1.3)

$$4\pi\mathbf{j} = (0, 0, -c\nabla^2 A). \tag{4.6}$$

Because of the symmetry $(\mathbf{j}.\nabla)\mathbf{H}$ is zero and the equilibrium condition expressed by (4.4) now becomes

$$(\mathbf{H}.\nabla)\,\mathbf{j}=0, \tag{4.7}$$

so that j is constant on a line of force (but it can have any value at a neutral point, since there (4.7) is automatically satisfied). Since A is constant on a line of force, this result may be expressed as

$$\nabla^2 A = F(A), \tag{4.8}$$

where F is an arbitrary function of A.

Equation (4.8) has been studied long ago because of its hydrodynamical applications (Lamb, § 165). An obvious solution is

$$A=\tfrac{1}{2}\alpha x^2+\beta xy+\tfrac{1}{2}\gamma y^2,$$

giving coaxial conics for the lines of force and

$$H^2=(\alpha x+\beta y)^2+(\beta x+\gamma y)^2,$$

which does not behave satisfactorily as the distance tends to infinity, however.

(i) *Models with property* (A). It was shown in § 3.2 that the property (A) was equivalent to the condition that Φ and hence also H^2 be constant on a line of force. Then $(\mathbf{H}.\nabla)\mathbf{H}$ is perpendicular to $\mathbf{H}$ and the discussion in § 4.1 of the curvature vector $\mathbf{k}$ shows that

$$(\mathbf{H}.\nabla)\mathbf{H}=H^2\mathbf{k}.$$

Then (4.5) requires $\qquad H^2\mathbf{k}=\nabla\Phi.$

For two-dimensional models $\mathbf{k}$ must be parallel to ∇A and consequently Φ is a function of A (this could be deduced from the condition that Φ is independent of z). Now, since $|\nabla A|=H$ and is constant on a line of force, we see that $|\mathbf{k}|$ must be constant on a line of force so that every line of force is circular. Next, since $|\nabla A|$ is constant on a line of force, the lines of force must be concentric circles (or parallel straight lines). These are, therefore, the *only* lines of force with two dimensional symmetry, that have the property (A). It may be noted that this argument cannot be generalized to three dimensions, because then $\mathbf{k}$ need not be parallel to ∇H^2.

When the lines of force are concentric circles, cylindrical coordinates r, θ and z are convenient; $H_r = H_z = 0$ and H_θ is an arbitrary function of r and is independent of θ and z. Then the total pressure $\Phi/4\pi$ is a function of r only, determined, apart from an additive constant, by

$$\frac{\partial \Phi}{\partial r} = -H_\theta^2/r, \tag{4.9}$$

which shows that Φ always decreases with increasing r. This illustrates the effect of the term $(\mathbf{H}.\nabla)\mathbf{H}$. Suppose that the field is confined between two cylindrical surfaces $r=a$ and $r=b$, with $a<b$, so that for $r<a$ or $r>b$, $p=\Phi/4\pi$. Then the gas pressure inside the inner cylinder must be greater than the pressure outside the outer cylinder; the gas in the inner cylinder is squeezed by the tension in the lines of force.

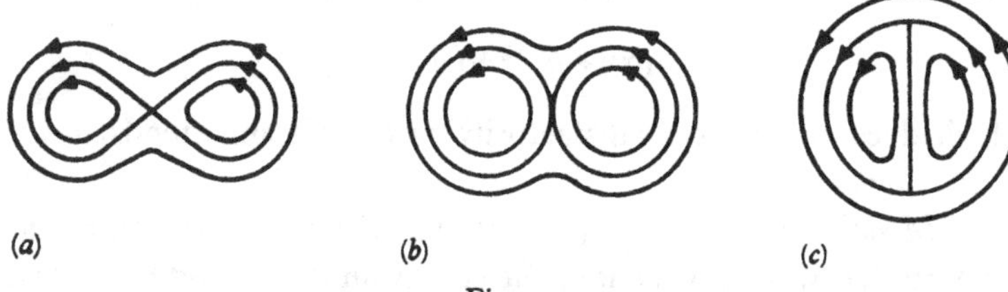

Fig. 4.2

It is obvious that any equilibrium model with just one neutral point has concentric circular lines of force, since the field energy is minimized by minimizing the lengths of the lines of force. Any other model must therefore have more than one neutral point. No static model is obtained with two neutral points; if there are two neutral points and the current densities at the neutral points are anti-parallel, the neutral points always move apart; if the current densities are parallel, there is a third (X-type) neutral point.

(ii) *Models with an X-type neutral point.* The next most complicated model has one X-type and two O-type neutral points, the topology of the lines of force being shown in fig. 42 (*a*); the limiting line of force through the X-type neutral point has the form of a figure-of-eight. The shape of the lines of force now depends on the relative strength of the field in different regions. If the field energy is concentrated in one narrow tube of force, this tube will take up a nearly circular shape, so as to minimize the energy. If the magnetic energy inside the loops greatly exceeds that outside, the lines of force inside will approximate to concentric circles and in the extreme case the figure of eight will consist of two circles in contact as in fig. 4.2(*b*). If the magnetic energy outside is much the greater, the lines of force outside will approximate to circles and in the extreme case the figure of eight will consist of two *D*'s back to back as in fig. 4.2(*c*). In any intermediate case the configuration will be intermediate between figs. 4.2(*b*) and (*c*), as shown in fig. 4.3. As

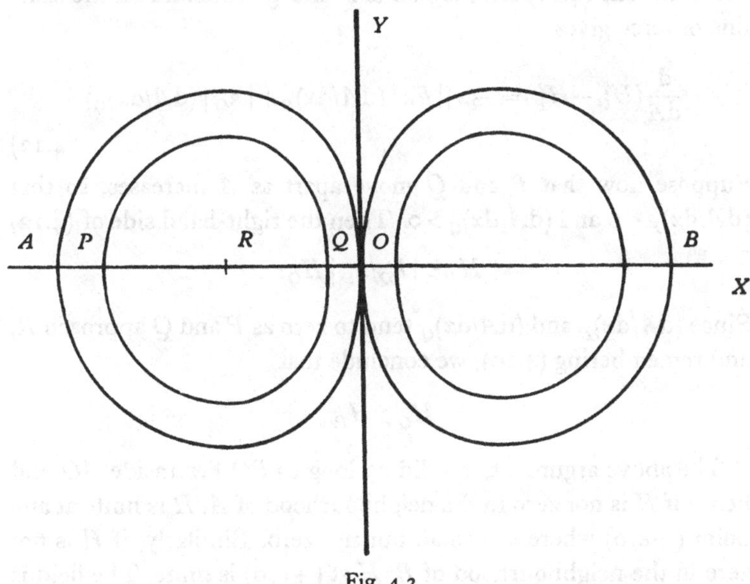

Fig. 4.3

in both (*b*) and (*c*) the angle between the limiting lines of force at the *X*-type neutral point is zero, this is also true in general. Also the field is symmetrical about the line through the three neutral points, which we now take as *x*-axis, the origin *O* being taken as the *X*-type neutral point. Consider a line of force inside one of the loops of the figure eight and cutting the *x*-axis in *P* and *Q* as in fig. 4.3 and let its curvature at *P* be k_P and at *Q*, k_Q. Now for fig. 4.2(*b*) $|k_P| = |k_Q|$ and for fig. 4.2(*c*) $|k_P| > |k_Q|$, so that in the general case

$$|k_P| \geqslant |k_Q|. \tag{4.10}$$

Let us now consider the variation of field strength along the x-axis using the vector potential $(0, 0, A)$ and remembering that p is a function of A. Using the representation with a curvature term and dividing the x-component of (4.3) by H, we obtain

$$k_x \frac{dA}{dx} = \frac{d}{dA}(4\pi p + \tfrac{1}{2}H^2) \tag{4.11}$$

the derivatives all referring to variations along the x-axis and the sign of k_x being positive at P and negative at Q. Now subtracting the equations (4.11) for the points P and Q which lie on the same line of force gives

$$\frac{d}{dA}(H_Q^2 - H_P^2) = -2\{|k_P|(dA/dx)_P + |k_Q|(dA/dx)_Q\}. \tag{4.12}$$

Suppose now that P and Q move apart as A increases, so that $(dA/dx)_P < 0$ and $(dA/dx)_Q > 0$. Then the right-hand side of (4.12) $\gtrless 0$ as

$$H_P \gtrless |k_Q/k_P| H_Q.$$

Since $(dA/dx)_P$ and $(dA/dx)_Q$ tend to zero as P and Q approach R, and remembering (4.10), we conclude that

$$H_Q \geqslant H_P.$$

The above argument, is valid so long as PQ lies inside AO and hence if H is not zero in the neighbourhood of A, H is finite at any point $(-\epsilon, 0)$ where ϵ is small but not zero. Similarly, if H is not zero in the neighbourhood of B, H at $(+\epsilon, 0)$ is finite. The field is directed in opposite directions at $(-\epsilon, 0)$ and $(+\epsilon, 0)$ and it follows that at **O** the field must change discontinuously and curl **H** must be infinite. This conclusion is required by the condition of mechanical equilibrium; its significance will be discussed in § 6.4.

(iii) *Models with a gravitational field.* When there is a gravitational field, the problem is much more complicated because the gravitational force is proportional to the mass density, so that its relation to the pressure involves the temperature; as a simplication the temperature is here taken to be uniform. The gravitational field is taken to be uniform and equal to $-g$ in the y-direction, which is therefore the upward direction. The vector potential and current density are $(0, 0, A)$ and $(0, 0, j)$ as before, and the x- and y-components of the static equation are

$$\frac{j}{c}\frac{\partial A}{\partial x}=\frac{\partial p}{\partial x}, \tag{4.13}$$

$$\frac{j}{c}\frac{\partial A}{\partial y}=\mu g+\frac{\partial p}{\partial y}. \tag{4.14}$$

It is convenient to write $\mu g=\alpha p$ with

$$\alpha=h^{-1}=mg/kT,$$

where m is the mean ionic mass; h is the scale height. Then, putting $q=p\,\mathrm{e}^{\alpha y}$, (4.14) may be written

$$\frac{j}{c}\frac{\partial A}{\partial y}=\mathrm{e}^{-\alpha y}\frac{\partial q}{\partial y}. \tag{4.15}$$

Using (4.13) it follows that q is constant on a line of force and hence that $j\,\mathrm{e}^{\alpha y}$ is constant on a line of force. This is a generalization of (4.7) and the corresponding generalization of (4.8) is

$$\nabla^2 A=\mathrm{e}^{-\alpha y}F(A).$$

A conformal transformation of co-ordinates

$$\xi+\mathrm{i}\eta=\exp\{-\tfrac{1}{2}\alpha(y+\mathrm{i}x)\} \tag{4.16}$$

gives
$$\frac{\partial^2 A}{\partial \xi^2}+\frac{\partial^2 A}{\partial \eta^2}=\mathrm{e}^{\alpha y}\nabla^2 A=F(A), \tag{4.17}$$

which has the same form as (4.8). Thus any solution of the magnetostatic problem in the absence of a gravitational field provides a solution for the problem in a uniform gravitational field with uniform temperature. Since ξ and η are both periodic in x with period $4\pi h$, all these solutions are similarly periodic; other solutions must of course exist.

The solutions corresponding to the concentric circles in (i) are of particular interest. Their lines of force are the lines of constant B where

$$B=\mathrm{e}^{-\alpha y}-2\,\mathrm{e}^{-\frac{1}{2}\alpha y}\cos\tfrac{1}{2}\alpha x+1 \tag{4.18}$$

the origin for x and y being arbitrary. They are shown in fig. 4.4: they are closed loops for $B<1$ and infinite wavy lines for $B>1$. The closed loops can be used to construct models which satisfy reasonable boundary conditions; the field can be made to vanish outside the cylinder obtained by projecting one of the closed loops in the z-direction. It may be noted that the total mass in this cylinder must be the same as it would be in the absence of a magnetic field, otherwise it would float upwards or sink. This condition is attained by the increase of total pressure inside the central loops. This kind of model appears to give a useful description of the coronal arches associated with quiescent prominences.

Fig. 4.4

4.4. Axially symmetric models

Using cylindrical co-ordinates r, θ and z, axially symmetric models are defined by the condition that H_r, H_θ and H_z are independent of θ; the gas pressure is then also independent of θ. It should be noted that the vanishing of $\operatorname{div} \mathbf{H}$ requires $H_r \to 0$ at least as fast as r, when $r \to 0$; this means that the lines of force cannot cross the axis. In the special case when $H_\theta = 0$ the lines of force lie in the meridian planes, but there are no realistic equilibrium configurations of this type without a gravitational field and none have been worked out with a gravitational field (but see Prendergast 1956). This can be seen from consideration of the energy. Suppose the field is confined within a finite region of space; if this region moves outward from the axis, any slice of it cut off by two meridian planes expands in the direction perpendicular to the meridian planes, and its magnetic energy is then reduced. Such a field will therefore move outward from the axis, unless it is held in by a gravitational field. It is possible that some stars have fields of this type, but no detailed model has been worked out for them.

It may be that, in certain regions, the current density is negligible; in such a region the equilibrium configuration of the gas is not affected by the magnetic field. The field cannot be source-free everywhere, but it may be useful to know the behaviour of a field in a particular region, where the sources are negligible. Axially symmetric source-free fields with $H_\theta = 0$ can all be expanded in spherical harmonics; the components which decrease with increasing distance from the axis have the well-known form of multipole fields. A less restrictive condition, which is more likely to be valid, is that the magnetic force density is negligible, in the sense that it is small compared with $H^2/4\pi ac$. This would be true, if the magnetic pressure greatly exceeded the gas pressure, and does not imply that the effect of the magnetic force on the gas is negligible; it probably is true in the solar atmosphere above sunspots. There is a much greater variety of axially symmetric fields with $\mathbf{j} \wedge \mathbf{H} = 0$ (and $H_\theta \neq 0$) than with $\mathbf{j} = 0$ and it is useful to discuss the general axially symmetric model satisfying (4.3). Such a model generally has linked lines of force and so avoids the kind of instability described in § 3.7. The problem may then be formulated by supposing that a field and distribution of material are given, not in mechanical equilibrium, and that the field is constrained to move with the material. The system will then have a configuration of minimum energy (provided that the material cannot be compressed indefinitely without absorbing all the field energy) and this will be stable. If the given configuration is axially symmetric, it may be expected that the equilibrium configuration is too, though it will appear later that this is not always the case.

A convenient representation, satisfying (3.1), in cylindrical co-ordinates, is

$$H_r = -\frac{1}{r}\frac{\partial B}{\partial z}, \quad H_z = \frac{1}{r}\frac{\partial B}{\partial r}.$$

Because of the symmetry B is constant on the surface of revolution generated by a line of force. Such surfaces can be described by the curves of constant B on a meridian plane; a line of force generally covers the whole of the surface. The components of $\mathbf{j}$ in cylindrical co-ordinates are given by

$$4\pi j_r/c = -\partial H_\theta/\partial z, \quad 4\pi j_z/c = \frac{1}{r}\frac{\partial}{\partial r}(rH_\theta),$$

$$4\pi j_\theta/c = -\frac{1}{r}\frac{\partial^2 B}{\partial z^2} - \frac{\partial}{\partial r}\frac{1}{r}\frac{\partial B}{\partial r},$$

using the axial symmetry. For equilibrium, the symmetry further requires that the θ-component of the force density should vanish, and hence

$$(\mathbf{H}.\nabla)(rH_\theta) = 0.$$

Thus rH_θ is constant on a line of force and so also by symmetry on the surface of revolution given by $B=$ constant, and we may write

$$H_\theta = C(B)/r.$$

Using the fact that the gas pressure p is also a function of B, the r- and z-components of (4.2) reduce to

$$\frac{\partial^2 B}{\partial z^2} + r\frac{\partial}{\partial r}\frac{1}{r}\frac{\partial B}{\partial r} = -\frac{1}{2}\frac{dC^2}{dB} - r^2\frac{dp}{dB}, \tag{4.19}$$

which is the general equation of equilibrium in the form corresponding to (4.8) for two-dimensional symmetry. As mentioned before, the solutions are restricted by the condition that the lines of force do not cross the axis $r=0$, and this requires B to be constant on the axis. This rules out many of the most obvious solutions. The special case when $\mathbf{j}\wedge\mathbf{H}$ vanishes is obtained by omitting the pressure term in (4.19).

4.5. Instability of models with helical lines of force

An obvious solution of (4.19) is obtained by taking B to be a function of r only; C can then be any function of r. For such a field, $H_r = 0$ and H_θ and H_z are arbitrary functions of r. Since H is constant on a line of force, the field has the property (A); it is simply the modification obtained by adding a field $H_z(r)$ to the two-dimensionally symmetric model whose lines of force are concentric circles. Since the total pressure $4\pi\Phi$ is unaffected by H_z, it decreases with increasing r as before; the lines of force are helical and squeeze the gas inside them. The pitch of the helices, which are the lines of force, is $2\pi P$, where

$$P = rH_z/H_\theta,$$

and is in general a function of r.

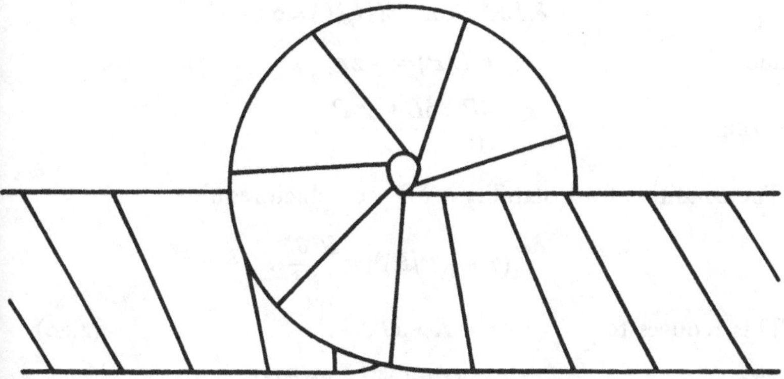

Fig. 4.5

This family of models is obviously the simplest with twisted lines of force. The models are rather trivial, but a certain type of instability occurring in them may be expected to occur in more general twisted fields and is very interesting. Consider a model in which the field vanishes outside a cylinder of radius R. Such a model is analogous to a twisted elastic string and it is common experience that when a string is twisted, there comes a stage when small loops form as shown in fig. 4·5. Alfvén (1950*b*) pointed out that this should happen to the magnetic lines of force as well, and suggested that it would lead to the generation of flux: in the case of a magnetic field, but not of a string, a loop can break away by the mechanism described in § 3.6. He estimated the critical condition for the formation of loops by considering the energy of a model with H_z and H_θ/r constant inside the cylinder, $r < R$. Suppose that a certain length L of the tube is involved; the magnetic energy contained in this length before the formation of a loop is given by

$$W = \tfrac{1}{4}L\int_0^R (H_z^2 + H_\theta^2)\, r\,\mathrm{d}r = \tfrac{1}{8}LR^2H_z^2(1 + \tfrac{1}{2}(R/P)^2).$$

The magnetic energy of the same part of the tube after the loop is formed may be estimated by considering the loop to be straightened out. The number of twists in this part of the tube is then one less than the original number, but the length is greater and may be written $L + \delta L$, where $\delta L \approx 2\pi R$. Supposing the volume to remain constant we may write

$$\delta(LR^2) = 0, \quad \delta(H_z/L) = 0$$

and

$$\delta(L/P) = -2\pi,$$

giving

$$\frac{\delta P}{P} = \frac{\delta L + 2\pi P}{L}.$$

The condition for instability is $\delta W < 0$, which requires

$$\frac{\delta L}{L}(2 + \tfrac{1}{2}(R/P)^2) < \frac{R^2\delta P}{P^3}.$$

This reduces to

$$R > \alpha P, \tag{4.20}$$

with

$$\alpha = (4 + (2\pi R/\delta L)^2)^{\frac{1}{2}} - 2\pi R/\delta L.$$

This result is independent of L and shows that α does not depend very much on the guessed value of $\delta L/2\pi R$. If $\delta L = 2\pi R$, $\alpha = \sqrt{5} - 1$ and in any case $\alpha < 2$. Lundquist (1951) made a more detailed investigation of the change of magnetic energy for a quite general field and applied it to helical fields in which H_z and H_θ were arbitrary functions of r. He concluded that instability usually occurs when the energy of twisting $(8\pi)^{-1}\int H_\theta^2 \mathrm{d}V$ is of the same order as the energy of the untwisted field $(8\pi)^{-1}\int H_z^2 \mathrm{d}V$. If the pitch and field strength vary greatly with r, however, there may be quite a difference in the behaviour of different models, both in the condition of critical stability and in the shape of the loops formed.

A treatment of the same problem, using Alfvén's particular model, by the method of normal modes (see § 5.6) was made by Dungey and Loughhead (1954). This showed that the model is already unstable with respect to other modes before it is twisted enough to form loops. One of these modes also occurs for a string and may be described by saying that the string as a whole takes a helical form (the material that is initially on the axis is displaced on to a helix). Equilibrium configurations with helical symmetry should therefore exist, but they are more complicated than axially symmetric models, and have not been worked out.

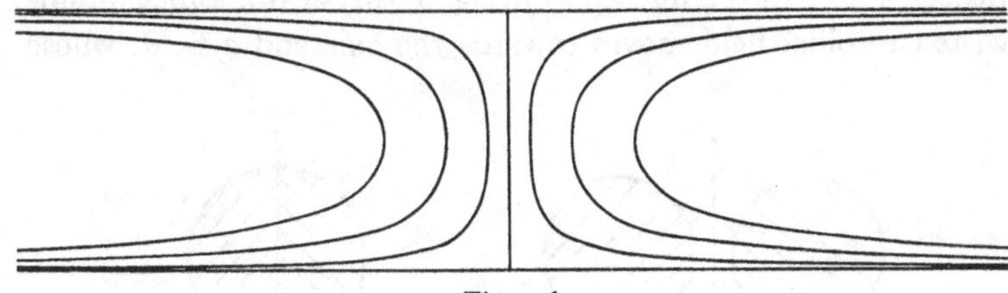

Fig. 4.6

Another mode of instability leads to an axially symmetric distortion; the radial displacement varies sinusoidally with z. This shows that twisting may cause narrow necks to form in the tubes of force. This effect is also illustrated by some simple solutions of (4.19). In these $H \to \infty$ as $r \to \infty$, but the qualitative discussion of the next section suggests that similar solutions exist in which the field is restricted to a finite volume in space. These solutions are given by

$$B = \beta r^2(\cos kz + \gamma),$$

with

$$C = \pm kB,$$

and

$$p = p_0 - \beta\gamma k^2 B.$$

Then

$$H_r = \beta kr \sin kz,$$

and

$$H_z = 2\beta(\cos kz + \gamma).$$

The model with $\gamma = 0$ provides an example of a field for which $\mathbf{j} \wedge \mathbf{H}$ vanishes. The model with $\gamma = 1$ is of interest, because, when $\cos kz = -1$, H_r, H_z, B, C and hence H all vanish. Any plane with $\cos kz = -1$ can then be used as a boundary, on one side of which there is no field. Thus the field can be confined to the region between two planes such as $z = \pm \pi/k$, as shown in fig. 4.6. The striking feature of this model is the way in which the lines of force splay out near the boundaries. This is a consequence of the twisting of the field.

4.6. Twisted toroidal fields

In none of the models yet discussed is the field confined to a finite volume, as it should be for a completely satisfactory model. The simplest satisfactory model is axially symmetric, and will now be discussed qualitatively in the light of the preceding results. The region containing the field takes the form of a doughnut with or without a hole in the middle as in fig. 4.7(*a*) or (*b*). Since an untwisted toroidal field moves towards the axis and a field, whose

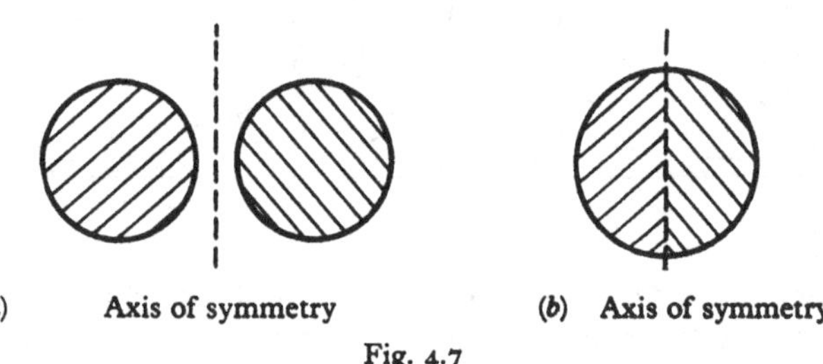

Fig. 4.7

lines of force lie in the meridian planes, moves away from the axis, the form (*a*) must occur for more strongly twisted toroidal fields than (*b*). Fields of this type must be formed, when a loop breaks away from a parent tube of force; considering the form of a loop when it is generated by twisting, it seems probable that it is near the condition of transition between (*a*) and (*b*). Though the degree of twisting in the loop in fig. 4.5 might differ considerably from that in the straight part of the tube, (4.20) gives a rough value for the degree of twisting required for this transition. By analogy with the formation of loops in a straight twisted tube of force, it may be expected that some configurations of type (*a*) are unstable and go into stable configurations of the type shown in fig. 4.8, which is not axially symmetric. If, as just suggested, the condition for the transition from (*b*) to (*a*) is the same as that for the formation of loops, it would be expected that any configuration of type (*a*) would be unstable in this way. The condition depends, however, on the way in which the degree of twisting varies with the distance from the axial line of force, and we next discuss another extreme case, which shows that stable configurations of type (*a*) must exist.

Consider the axially symmetric model shown in fig. 4.9, where the field is confined to the regions U and V and the field in U is untwisted ($H_r = H_z = 0$). Given the field in V, the field in U is determined by the equations

$$\frac{\partial \Phi}{\partial r} = -\frac{H_\theta^2}{r}, \quad \frac{\partial \Phi}{\partial z} = 0,$$

and hence
$$\frac{\partial H_\theta}{\partial z} = 0.$$

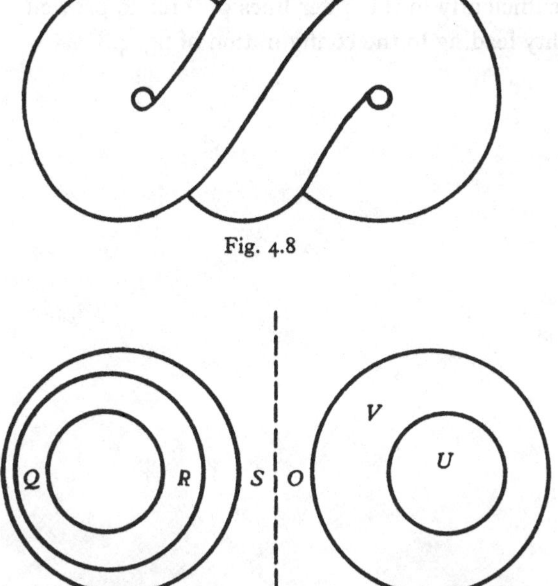

Fig. 4.8

Axis of symmetry

Fig. 4.9

The value of Φ on the inner boundary of V must therefore decrease with increasing r, but is not otherwise restricted. Since the field vanishes outside V, Φ must be constant on the outer boundary of V. To see that these conditions can be satisfied, suppose that the outer boundary of V is circular and that, in V, $H_\theta = 0$. In fig. 4.9 the outer boundary of V is the circle with PS as diameter and the curve through Q and R is a line of force just inside this boundary. Since H is constant on the outer boundary RS exceeds PQ by the factor OP/OS. Now the curvature of the line of force is $2/PS$ at both P and S and therefore

$$\Phi_R - \Phi_Q = \frac{2H^2}{PS}(RS - PQ)$$

which is positive. Stable configurations of type (*a*) therefore exist, and our general conclusion is that two requirements must be fulfilled. There must be sufficient twisting to prevent the central hole from closing, as in fig. 4.7(*b*). The twisting must be concentrated sufficiently in the outer lines of force to prevent the type of instability leading to the configuration of fig. 4.8.

CHAPTER 5
MAGNETODYNAMICS

5.1. General remarks

In the most important problems of cosmic electrodynamics the material is in motion and the magnetic field varies with time. In these problems the results of chapter 3 are needed as well as the effect of the magnetic force density on the motion. The combination of these effects is often too complicated to be represented by mathematical models and so much of the discussion is descriptive. It was seen in chapter 3 that it is nearly always permissible to neglect the motion of the magnetic field relative to the material. Then the picture of the lines of force as elastic strings is very useful in the descriptive work, because it incorporates both the motion of the field and the magnetic force density. An important general conclusion obtained from this picture is that motion occurring in any particular region is likely to influence the motion in any other region which is connected to it by lines of force. For this reason it is important to have some idea of the configuration of the field; often important conclusions can be reached without having a detailed model, but not much can be deduced from the strength of the field without any knowledge of its morphology.

5.2. Models with steady flow

Some steady-state models are not too complicated for a mathematical discussion. To obtain models with a steady flow and a constant magnetic field it is necessary to neglect all dissipative effects. We therefore take the viscosity to be zero and the conductivity to be infinite. We also neglect any motion of the field relative to the material due to other causes, as they are usually unimportant, and for simplicity we suppose that there is no gravitational field. Then the equations for a steady state are

$$\operatorname{curl}(\mathbf{u} \wedge \mathbf{H}) = 0 \tag{5.1}$$

and

$$4\pi\mu(\mathbf{u} \cdot \nabla)\mathbf{u} = (\mathbf{H} \cdot \nabla)\mathbf{H} - \nabla\Phi, \tag{5.2}$$

which requires

$$\operatorname{curl}(\mu(\mathbf{u} \cdot \nabla)\mathbf{u}) = \operatorname{curl}((\mathbf{H} \cdot \nabla)\mathbf{H}/4\pi). \tag{5.3}$$

In this section we discuss only models in which the motion is everywhere parallel to the field; then (5.1) is automatically satisfied. A simple solution of (5.3) occurs when μ is constant on a line of force and is given by

$$(4\pi\mu)^{\frac{1}{2}}\mathbf{u}=\mathbf{H}. \tag{5.4}$$

This solution exists for an arbitrary field $\mathbf{H}$. The inertial force in (5.2) exactly balances the 'curvature' part of the magnetic force and Φ is just a constant. The lines of force may be pictured as thin-walled elastic tubes with water running through them; when the water goes round a curve in the tube the centrifugal force keeps the tube stretched.

The models described by (5.4) have a certain stability. Consider a tube of force with material circulating parallel to the field, but suppose that (5.4) does not hold initially. The tube will then expand or contract in length depending on whether u is larger or smaller than the value given by (5.4). Now Thomson's theorem, which is the hydrodynamic analogue of the 'freezing-in' theorem, states that $\oint \mathbf{u}.d\mathbf{s}$, taken round any closed curve moving with the material, is constant for a model with zero viscosity. Consequently, stretching the tube causes a decrease in u and an increase in H and vice versa. If then initially $\mathbf{u}$ is parallel to $\mathbf{H}$, but does not satisfy (5.4), the tube will expand or contract towards the steady state condition and in this sense the steady state is stable.

The decay due to viscosity of the motion given by (5.4) is similar to the decay of the field due to Ohmic loss and both are usually slow in astrophysical systems in the absence of turbulence. One might well expect a flow of material along the lines of force to occur sometimes in magnetic fields on the sun, and such a flow would be long-lived. To describe such a phenomenon the models discussed in this section would need to be modified by the inclusion of the gravitational field. This will not be pursued, because, rather surprisingly, such models do not fit any of the phenomena observed at all obviously.

5.3. Rotating fields

Models with steady rotation are of great importance in cosmic electrodynamics; they can be applied to the rotation of stars and possibly planets. The problem is very much simplified by neglecting the motion of the field relative to the material. It may then be noted immediately that: (i) if the field is symmetric with respect to the axis of rotation, it is constant in time; and (ii) if the rotation is uniform, any field just rotates without suffering any distortion.

It is necessary to consider non-uniform rotation, because, for instance, the observed solar rotation is non-uniform. There is no motion in the meridian planes, so that $u_r = u_z = 0$ and we may write

$$\mathbf{u} = \mathbf{R} \wedge \boldsymbol{\omega}, \tag{5.5}$$

where $\mathbf{R}$ is the position vector from an origin on the axis and $\boldsymbol{\omega}$ is parallel to the axis, but $|\boldsymbol{\omega}|$ may vary with r and z, though not with θ. Now suppose that ω is larger at some point P than at another point Q on the same line of force. Then in a given time P will have rotated through a larger angle than Q and after a long enough time P will have made many more complete rotations than Q. Thus, the line of force through P and Q is continually wound round the axis of rotation. In this case the system is not in a steady state and obviously the tension in the lines of force resists the winding motion and so tends to make the velocities at P and Q equal. In general this results in oscillations, which will be described in § 5.5, but the conclusion required here is that in a steady state ω must be constant on any line of force. This result was first obtained by Ferraro (1937) who called it 'the law of isorotation'; it is strictly true for a steady state and nearly true in general over a time long enough for short-lived disturbances to be neglected.

The law of isorotation is generally discussed in relation to fields which are symmetric about the axis of rotation. The situation is similar if the lines of force have this symmetry but the field strength varies with θ. If the lines of force did not have this symmetry, the law of isorotation, together with the condition that ω is independent of θ, would require the rotation to be uniform throughout the region occupied by the field. If then the time-averaged field can be defined so that it moves with the material (disturbances in the field might result in motion of the mean field relative to the material, as in § 5.8) its lines of force must be axially symmetric, if the rotation is non-uniform.

Consider now the forces acting in a rotating star with a magnetic field; with axial symmetry their directions lie in the meridian plane.

The inertial force density $-\mu(\mathbf{u}.\nabla)\mathbf{u}$ reduces, using (5.5), to just the centrifugal force $\mu\omega^2 r$ in the radial direction. Now assuming that the surrounding gas is ionized the law of isorotation requires all the surrounding gas contained in tubes of force from a rotating star to rotate with the star and, if this rotating gas extends far from the axis of rotation, the centrifugal force becomes large. If the lines of force do not extend far from the axis of rotation, the centrifugal force might not be very important, but this is not the case for the dipole field for instance. It is therefore important to consider the region swept out by the rotating lines of force and the conditions on the boundary of this region.

If a rotating star (or planet) had no magnetic field, the surrounding gas would tend to rotate with the star, but the rate of rotation would be controlled by the viscosity and ω would decrease gradually with the distance from the star. When there is a magnetic field, on the other hand, the rate of rotation is controlled by the law of isorotation. The viscous retardation acting on the outlying gas is communicated to the main body of the star by the magnetic field; the rate of retardation $\partial\omega/\partial t$ must be constant on a line of force and, because a star has a very large mass, must be very small. Thus the gas in the surrounding region rotates with the same angular velocity as the star, this surrounding region being defined by the lines of force which emerge from the star. If this region were to extend far from the axis, the centrifugal force would become very large, and there must evidently be some limitation resulting from this consideration. Consider the equatorial plane; beyond a certain distance r_0 the centrifugal force exceeds the force due to gravity. Now, since the magnetic force density is perpendicular to the field, the variation of pressure along a line of force is controlled by the centrifugal and gravitational forces only, and beyond the critical distance the pressure must increase with r. But in the region surrounding a star such an increase cannot continue indefinitely and probably the pressure should not become larger than that of the interstellar gas. Thus the lines of force must not go farther from the axis than a certain limit, which is probably not much greater than r_0. Fig. 5.1 shows an axially symmetric model which satisfies this requirement; it contains neutral points N. There may be a field outside the outer line of fig. 5.1; if so, this outer field can move independently of the field shown. If the rotation is non-uniform, the value of r_0 depends on the value of ω for the outer lines, which are those emerging from the stellar surface at the poles. The gas outside the outside line of force rotates more slowly and there is a viscous force acting on the outer lines. This winds up the outer

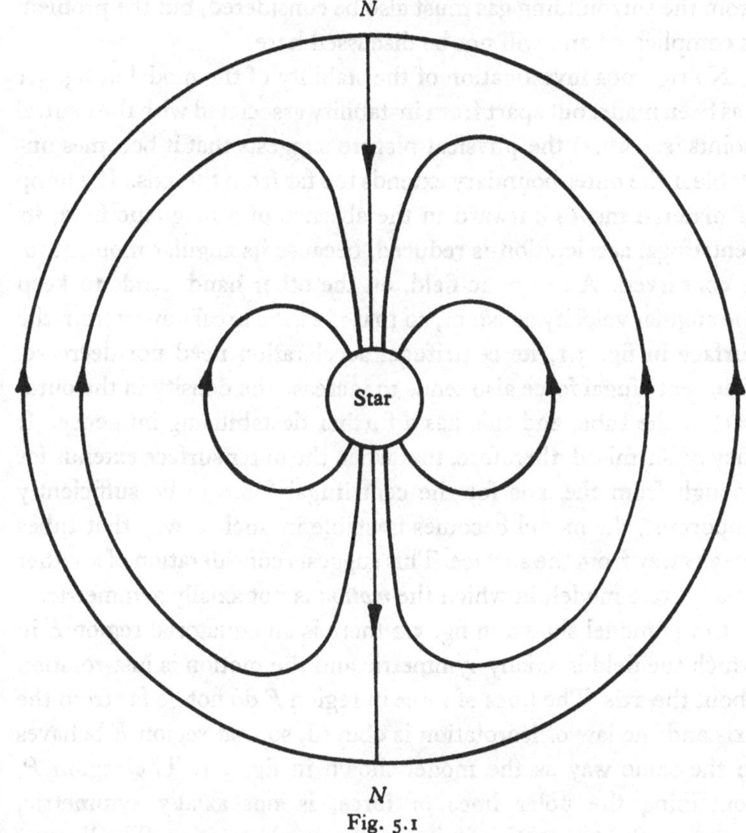

Fig. 5.1

lines of force and so communicates the viscous retardation to the core of the star (the core being a tube near the axis and containing the polar lines of force). For the model of fig. 5.1 this retardation of the core of the star will cause a decrease of angular velocity with increasing latitude at the surface of the star, as is observed on the sun.

In general, currents must flow in the stellar atmosphere and the surrounding gas, and, over a long time, the motion of the field relative to the material may be appreciable. The resulting change of the field is likely to be compensated by a suitable motion of the material, however; the Hall effect requires a slight modification of the law of isorotation and the other effects can be compensated by motion in the meridian planes. The accretion by the star of material from the surrounding gas must also be considered, but the problem is complicated and will not be discussed here.

No rigorous investigation of the stability of the model of fig. 5.1 has been made, but apart from instability associated with the neutral points (see §6.4) the physical picture suggests that it becomes unstable if the outer boundary extends too far from the axis. If a lump of material moves outward in the absence of a magnetic field, its centrifugal acceleration is reduced, because its angular momentum is conserved. A magnetic field, on the other hand, tends to keep the angular velocity constant, so that if a tube breaks away from the surface in fig. 5.1, its centrifugal acceleration need not decrease. The centrifugal force also tends to increase the density in the outer part of the tube, and this has a further destabilizing influence. It may be surmised, therefore, that when the outer surface extends far enough from the axis for the centrifugal force to be sufficiently important, the model becomes unstable in such a way that tubes break away from the surface. This suggests consideration of another steady state model, in which the *motion* is not axially symmetric.

In the model shown in fig. 5.2 there is an equatorial region E in which the field is axially symmetric and the motion is just rotation about the axis. The lines of force in region E do not go far from the axis and the law of isorotation is obeyed, so that region E behaves in the same way as the model shown in fig. 5.1. The region P, containing the polar lines of force, is not axially symmetric, however, and the motion in it is not a simple rotation. The lines of force in region P may go far from the axis (fig. 5.2 is only schematic) and the motion must be such that the centrifugal force is not too large. This can be achieved if region P is a closed tube twiddling about its own axis. The tube P can have any form provided its radius is not too large; such a tube cannot be axially symmetric and in the simplest model is symmetrical with respect to a particular meridian plane (in fig. 5.2 the plane of the paper). Now the lines of force in this tube can rotate about the tube's own axis (shown dotted in fig. 5.2) for then neither will the rotation twist the lines of force nor will the centrifugal force be too large.

Again, for the model of fig. 5.2 there will be a viscous force acting on the outer lines which will tend to twist the lines of force, and,

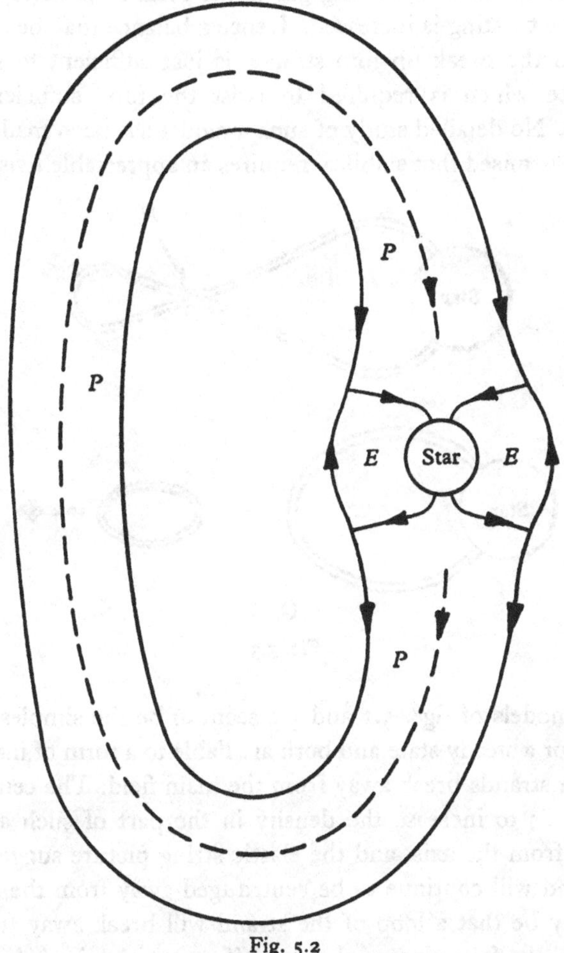

Fig. 5.2

if the field extends into H I regions, the neutral particles contribute to the retardation of the tube's rotation. Although the centrifugal force can be reduced by making the tube narrow, this model may be unstable in a somewhat similar way to the model of fig. 5.1, and small 'strands' may tend to break away from the main tube.

The occurrence of this type of instability will be demonstrated in § 5.6, but has not been studied in detail. The elastic tube picture suggests that twisting of the tube would tend to keep the strands together and so stabilize the model. If strands do break away the resistance of the surrounding gas to the rotation is increased and hence the twisting is increased. Hence a balance may be achieved in which the break up into strands is just sufficient to give the resistance which is required to twist the tube sufficiently for stability. No detailed study of such a model has been made, but it may be surmised that stability requires an appreciable twist.

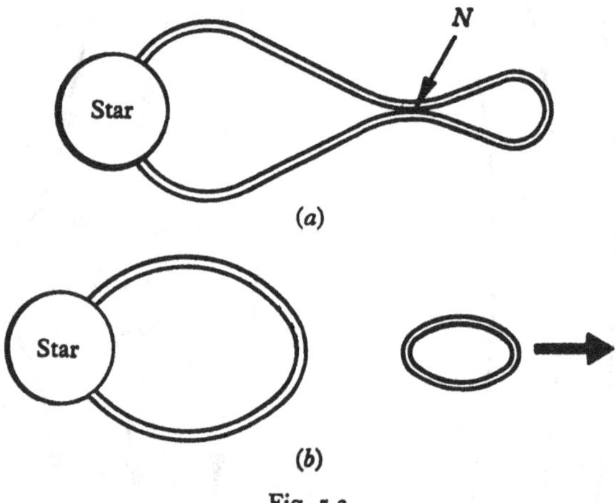

Fig. 5.3

The models of figs. 5.1 and 5.2 seem to be the simplest possibilities for a steady state and both are liable to a form of instability in which strands break away from the main field. The centrifugal force tends to increase the density in the part of such a strand farthest from the axis, and the elastic string picture suggests that the strand will continue to be centrifuged away from the axis. If so it may be that a loop of the strand will break away from the strand by the following mechanism. If a neutral point is formed as at N in fig. 5.3*a* the investigation to be described in § 6.4 shows that $\mathbf{j}$ can become very large there, and the Ohmic electric field then causes the outer loop to break away as in fig. 5.3*b*. If this occurs the strand loses both mass and angular momentum, but since the angular momentum per unit mass is greater for the loop which breaks away than for the whole strand, the angular momentum per unit mass of the strand left behind is reduced. Consequently, many repetitions of this phenomenon, though not enough to reduce appreciably the mass of the strand (including the part in the stellar interior) may substantially reduce its angular velocity. Thus the models of figs. 5.1, 5.2 and 5.3 all suggest that the rotation of that part of the star connected to the outer lines of force in the surrounding gas, is retarded. An attempt at a rough quantitative estimate of this effect will be made for the sun in §7.5 (see also Lüst and Schlüter, 1955).

5.4. Alfvén waves

Some hydrodynamic and electromagnetic phenomena are most simply described in terms of waves, and the wave description is useful in cosmic electrodynamics. Here Alfvén waves are involved: picturing the lines of force as elastic strings they are analogous to waves in real elastic strings. Under laboratory conditions they are heavily attenuated by Ohmic losses and they were discovered by theoretical means by Professor H. Alfvén, who called them magnetohydrodynamic waves. They have since been demonstrated in the laboratory using mercury (Lundquist, 1949).

In an ionized gas there is a continuous transition from radio waves to Alfvén waves as the frequency decreases, and it is here assumed that the frequency is low enough for the approximations made in § 1.3 to be valid. Waves travelling at a general angle to $\mathbf{H}$ are of a hybrid type between Alfvén waves and sound waves. We first discuss pure Alfvén waves travelling in the direction of the field; these are transverse. Only waves in uniform fields will be discussed, and at first dissipative effects and the Hall effect will be ignored. The field then moves with the material and the force is just the magnetic force. In this approximation the only physical effects included are those of electromagnetic induction.

Let $\mathbf{H}_0$ be a uniform field in the z-direction and let

$$\mathbf{H}=\mathbf{H}_0+\mathbf{h},$$

where $\mathbf{h}$ is small and perpendicular to $\mathbf{H}_0$. Let $\mathbf{u}$ be small and perpendicular to $\mathbf{H}_0$ and let the material be uniform, so that μ is a constant, and incompressible, so that div $\mathbf{u} = 0$. As usual in the theory of waves, all terms which are non-linear in $\mathbf{h}$ and $\mathbf{u}$ will be neglected, but note that no special form for the spatial or temporal variation of $\mathbf{h}$ and $\mathbf{u}$ is assumed. Then from (3.4)

$$\partial \mathbf{h}/\partial t = (\mathbf{H}_0 . \nabla)\, \mathbf{u}. \tag{5.6}$$

This provides a specially simple example of the freezing-in of the field. If the displacement of the material is $\boldsymbol{\xi}$, then $\mathbf{u} = \partial \boldsymbol{\xi}/\partial t$ and (5.6) can be integrated to give

$$\mathbf{h} = (\mathbf{H}_0 . \nabla)\, \boldsymbol{\xi}, \tag{5.7}$$

since $\mathbf{h}$ vanishes for the undisplaced field.

Using (4.1) the acceleration of the material is given by

$$4\pi\mu\, \partial \mathbf{u}/\partial t = (\mathbf{H}_0 . \nabla)\, \mathbf{h} \tag{5.8}$$

and hence
$$\left(4\pi\mu \frac{\partial^2}{\partial t^2} - H_0^2 \frac{\partial^2}{\partial z^2}\right)(\boldsymbol{\xi}, \mathbf{u}, \mathbf{h}) = 0. \tag{5.9}$$

This is a wave equation with the constant wave velocity V_A given by

$$V_A = (4\pi\mu)^{-\frac{1}{2}} H_0.$$

Suppose now that $\boldsymbol{\xi}$ is given initially: $\boldsymbol{\xi}$ is restricted only to be perpendicular to $\mathbf{H}_0$ and to be solenoidal. Then one solution of (5.9) is given by

$$\partial \boldsymbol{\xi}/\partial t = V_A\, \partial \boldsymbol{\xi}/\partial z,$$

and $\mathbf{h}$ is determined by (5.7); also for this solution

$$\mathbf{h} = (4\pi\mu)^{\frac{1}{2}}\, \mathbf{u}. \tag{5.10}$$

In this solution the disturbance represented by $\boldsymbol{\xi}$ simply moves as a whole with the velocity V_A in the z-direction without suffering any deformation. There is another solution representing a wave travelling in the opposite direction and the general solution is a superposition of two waves travelling in opposite directions.

It may be noted that the solutions representing waves travelling in one direction only are valid for arbitrarily large amplitude. They may be obtained from steady states satisfying (5.4) by transforming to a frame with velocity $(4\pi\mu)^{-\frac{1}{2}}\mathbf{H}_0$ in the frame of the steady state.

5.41. Effect of the Hall field on Alfvén waves

Alfvén waves provide a useful illustration of the effect of the Hall field. When this is included, but dissipative effects are still ignored, (5.6) becomes

$$\partial \mathbf{h}/\partial t = (\mathbf{H}_0 . \nabla)(\mathbf{u} - c\,\mathrm{curl}\,\mathbf{h}/4\pi n e). \tag{5.11}$$

Since the Hall effect causes dispersion it is best to consider a sinusoidal plane wave with

$$\mathbf{u}, \mathbf{h} \propto \exp[\mathrm{i}(\omega t - kz)], \tag{5.12}$$

where the frequency ω is real and the wave number k is real for an unattenuated wave. With the Hall effect included the normal modes are circularly polarized with $u_x = \pm \mathrm{i} u_y$ and $h_x = \pm \mathrm{i} h_y$. The dispersion equation derived from (5.11) and (5.8) may be written

$$\omega^2 = k^2(V_A^2 \pm \omega c H_0/4\pi n e). \tag{5.13}$$

This shows that the importance of the Hall effect increases with increasing frequency. At a given frequency it is most important for *weak* fields since $V_A \propto H_0$. The value of ω which makes the right-hand side of (5.13) vanish is approximately $\mp \Omega_p$. In most cosmic problems $|\omega| \ll \Omega_p$ and the Hall effect is not very important. Since it causes a difference in the wave velocities of the two circularly polarized modes of equal frequency it has the effect of rotating the plane of polarization of a plane polarized wave.

One way of describing the Hall effect is to say that the protons are less strongly frozen to the field than the electrons, and this may be illustrated for Alfvén waves. If the $\pm\omega$ in (5.13) is negative, the direction of rotation of the wave is the same as that of positive particles gyrating in the field $\mathbf{H}_0$. Now for any short wavelength $k \gg \Omega_p/V_A$ $(= (4\pi n e^2/m_p c^2)^{\frac{1}{2}}$ the 'proton plasma wave number') there is a root of (5.13) $\omega \approx \mp \Omega_p$. Thus for any disturbance varying rapidly in space the circularly polarized component corresponding to the direction of gyration of protons varies approximately sinusoidally in time with angular frequency Ω_p. The fundamental equations show that for such waves $E \ll |\mathbf{u} \wedge \mathbf{H}|/c$ and $\mathbf{u} \approx \mathbf{j}/ne$, so that the oscillations of the electrons and of the field are weak compared to those of the protons. Roughly speaking then the protons are free to perform a bulk gyration with their own gyroperiod without much disturbing the field. It may be noted that (5.13) is unsymmetrical between electrons and protons, but a more accurate treatment (Dungey, 1954) results in an additional factor which restores the symmetry. This factor is unimportant when $\omega<\Omega_p$, and the next section shows that the attenuation is usually large for $\omega>\Omega_p$.

5.42. Attenuation of Alfvén waves

The attenuation of Alfvén waves is caused by Ohmic losses and by viscosity. We assume here that the collision frequency is large enough for the usual treatment of the conductivity and the coefficient of viscosity to be valid; the case when the collision frequency is not so large has been discussed by the author (Dungey, 1954). Equations (5.6) and (5.8) now become

$$\partial\mathbf{h}/\partial t=(H_0.\nabla)\,\mathbf{u}+c^2\nabla^2\mathbf{h}/4\pi\sigma, \tag{5.14}$$

$$\partial\mathbf{u}/\partial t=(H_0.\nabla)\,\mathbf{h}/4\pi\mu+\eta\nabla^2\mathbf{u}. \tag{5.15}$$

Considering again sinusoidal plane waves defined by (5.12), and taking ω to be real, the imaginary part of k provides a measure of the attenuation. The dispersion equation obtained from (5.14) and (5.15) is

$$k^2V_A=(\omega-ic^2k^2/4\pi\sigma)(\omega-ik^2\eta). \tag{5.17}$$

Assuming the imaginary part of k to be much smaller than the real part we have

$$\frac{\mathscr{I}(k)}{\mathscr{R}(k)}\approx\frac{\omega}{2V_A^2}\left(\frac{c^2}{4\pi\sigma}+\eta\right), \tag{5.18}$$

which shows that the attenuation per wavelength is proportional to ω. Then putting $\sigma=ne^2/m_e\nu_{ep}$ (§2.7), $\eta=5kT/m_p\nu_{pp}$ (§2.8), $\nu_{pp}=(2m_e/m_p)^{\frac{1}{2}}\nu_{ep}$ (§2.5) and $2nkT=p$, (5.18) can be written

$$\frac{\mathscr{I}(k)}{\mathscr{R}(k)}=\left(\frac{5}{16(2m_p/m_e)^{\frac{1}{2}}}\right)^{\frac{1}{2}}\frac{\omega}{\Omega_p}\frac{(8\pi p)^{\frac{1}{2}}}{H}(\zeta+\zeta^{-1}), \tag{5.19}$$

where
$$\zeta=(4\pi\sigma\eta)^{\frac{1}{2}}/c.$$

Equation (5.19) shows that when ζ is varied the attenuation is a minimum for $\zeta=1$, and that, if $H^2<8\pi p$, the attenuation is always appreciable for $\omega>\Omega_p$. Using (from §2.7)

$$\nu_{ep}=\frac{(8\pi)^{\frac{1}{2}}}{3}\frac{e^4nA_1(2)}{m_e^{\frac{1}{2}}(kT)^{\frac{3}{2}}}$$

gives
$$\zeta=\frac{3(kT)^2}{e^3cA_1(2)}\left(\frac{5}{2n(2m_em_p)^{\frac{1}{2}}}\right)^{\frac{1}{2}},$$

which with $A_1(2)=30$, is $4.10^{-3}n^{-\frac{1}{2}}T^2$. When $\zeta > 1$ the viscous attenuation is more important than the Ohmic. At 1000° K $\zeta > 1$ for $n < 1{\cdot}6.10^7$ cm.$^{-3}$ and at 10^6 ° K for $n < 1{\cdot}6.10^{19}$ cm.$^{-3}$.

5·43. Waves in a compressible gas in a uniform magnetic field

So far only waves travelling in the direction of the field have been considered. Before discussing waves travelling in a general direction we must consider waves travelling perpendicular to **H**. These can only be longitudinal waves, and are similar to sound waves, but their velocity is increased by the magnetic field. When the material is compressed, **H** is increased, so that there is an additional restoring force resulting from the magnetic pressure, but there is no increase in inertia due to the field. The velocity of sound V_S in the absence of a magnetic field is given by

$$V_S = (\gamma p/\mu)^{\frac{1}{2}}, \tag{5.20}$$

where γ is the adiabatic value of $\partial(\log p)/\partial(\log \mu)$ and is $\frac{5}{3}$ for ionized hydrogen. Since the magnetic pressure $H^2/8\pi$ is proportional to μ^2 for longitudinal waves travelling perpendicular to the field, the effective value of γ for the magnetic pressure is 2. The wave velocity V_S' is then given by

$$V_S'^2 = V_S^2 + H^2/4\pi\mu = V_S^2 + V_A^2. \tag{5.21}$$

It should be noted that $(V_A/V_S)^2 = H^2/4\pi\gamma p$; when $H^2 \sim 8\pi p$, then $V_A \sim V_S$. The velocity of sound waves travelling parallel to the field is just V_S.

The results will now be quoted for waves whose wave normal has an arbitrary direction, inclined at an angle θ to the direction of **H**. There is one purely transverse mode, for which **u** and **h** are perpendicular to **H**; this has the wave velocity $V_A \cos\theta$, and is just a special case of the waves described by (5·9), since there **h** and **u** were allowed to vary with x and y. The other modes are in general neither purely transverse nor purely longitudinal. Their velocities are the values of V given by

$$V^4 - V^2(V_S^2 + V_A^2) + V_S^2 V_A^2 \cos^2\theta = 0.$$

When $V_A^2 \ll V_S^2$, one value of V^2 is approximately V_S^2 and the waves with this velocity are predominantly longitudinal. The other value is approximately $V_A^2 \cos^2\theta$ and the corresponding waves are predominantly transverse.

In the rare cases where $H_0^2 \gg 8\pi p$, so that the pressure gradient can be neglected, motion parallel to **H** is not involved and the general wave equation for Alfvén waves can be written in the simple-looking form

$$4\pi\mu\,\partial^2\mathbf{E}/\partial t^2 = -\mathbf{H}_0 \wedge (\text{curl curl}\,\mathbf{E}) \wedge \mathbf{H}_0 + [(\text{curl}\,\mathbf{H}_0) \wedge (\text{curl}\,\mathbf{E})] \wedge \mathbf{H}_0 \qquad (5.22)$$

and, to be compatible with the condition $H_0^2 \gg 8\pi p$, curl $\mathbf{H}_0$ must be nearly parallel to $\mathbf{H}_0$.

5.5. Small oscillations

A cosmic system with a magnetic field may have resonant modes of oscillation, which are equivalent to standing waves of the species described in the last section. A star already has resonant modes without a magnetic field, and the periods of the most fundamental of these can only be slightly altered by fields of the strength observed in stars, but the field also introduces additional modes. In the simplest model the field is axially symmetric. These additional modes are then torsional oscillations, which means that **u** and **h** are perpendicular to the meridian planes, and are also axially symmetric. Then the restoring force is entirely magnetic; it is just the force resulting from the winding up of the field, which was seen in § 5.3 to result when the law of isorotation is broken. With the dissipative terms omitted the wave equation for angular frequency ω can be written in cylindrical co-ordinates

$$4\pi\mu\omega^2 u_\theta + r^{-1}(\mathbf{H}_0 . \nabla)\, r^2(\mathbf{H}_0 . \nabla)(r^{-1}u_\theta) = 0. \qquad (5.23)$$

Since the only differential operator occurring here is $(\mathbf{H}_0 . \nabla)$, it follows that each surface of revolution of a line of force oscillates independently. This is an idealization resulting from the neglect of dissipative effects. Both the viscosity and Ohmic field couple the motions of adjacent lines of force but, provided these are small, the disturbance may be concentrated in a thin shell, and may be adequately described by (5.23) apart from the damping. Since the numerical integration of (5.23) along a line of force involves only one independent variable, it presents no computational difficulties.

The computation of other modes is more difficult. The W.K.B. approximation is usually valid only for high-order modes (those with many nodal surfaces), but it may be possible to estimate the period as the time of travel of a travelling Alfvén wave along a particular route. It is usually true that the period depends most on the region where the wave velocity is smallest, and that the energy density of the wave is largest there.

If the energy density can be put in a certain form the Rayleigh–Ritz method can be used to obtain an approximate value for the period. Representing the oscillation by the displacement $\boldsymbol{\xi}\cos\omega t$ the kinetic energy averaged over a cycle is

$$\omega^2 I_1 = \tfrac{1}{4}\omega^2 \int \mu\xi^2 \, dV,$$

and the method requires that the average potential energy I_2 should be an integral, bilinear in $\boldsymbol{\xi}$, such that a small variation $\delta\boldsymbol{\xi}$ changes I_2 by

$$\delta I_2 = -\frac{1}{2}\int \delta\boldsymbol{\xi}.\mathbf{F}\, dV, \tag{5.24}$$

where $\mathbf{F}$ is the force density. Then

$$\delta(\omega^2 I_1 - I_2) = \frac{1}{2}\int \delta\boldsymbol{\xi}.(\mu\omega^2\boldsymbol{\xi} + \mathbf{F})\, dV$$

and, since the acceleration is $-\omega^2\boldsymbol{\xi}$, this vanishes when $\boldsymbol{\xi}$ is the correct wave function. Now the average kinetic and potential energies are equal (this is Rayleigh's principle), and it then follows that I_2/I_1 is stationary for the correct wave function $\boldsymbol{\xi}$. Then the value of I_2/I_1 for a slightly incorrect $\boldsymbol{\xi}$ differs from ω^2 only in the second order in the error in $\boldsymbol{\xi}$. The Rayleigh–Ritz method is to assume a particular algebraic form for $\boldsymbol{\xi}$, with a few variable parameters, whose values are then chosen to make I_2/I_1 stationary. The value of I_2/I_1 is then the approximate value of ω^2.

The effect of a displacement $\boldsymbol{\xi}$ on the field $\mathbf{H}_0$ is determined by (3.4). Integration to first order in $\boldsymbol{\xi}$ gives

$$\mathbf{H} = \mathbf{H}_0 + \operatorname{curl}(\boldsymbol{\xi}\wedge\mathbf{H}_0) + O(\xi^2). \tag{5.25}$$

The first-order part of $\mathbf{F}$ is then

$$(4\pi)^{-1}\{[\operatorname{curl}\operatorname{curl}(\boldsymbol{\xi}\wedge\mathbf{H}_0)]\wedge\mathbf{H}_0 + (\operatorname{curl}\mathbf{H}_0)\wedge\operatorname{curl}(\boldsymbol{\xi}\wedge\mathbf{H}_0)\}.$$

Substitution of this in the right-hand side of (5.24) should give δI_2 and I_2 should be the second order part of the average potential energy, that is the second order part of $(8\pi)^{-1}\int \overline{H^2}\,\mathrm{d}V$. By partial integration, assuming that $\boldsymbol{\xi}$ vanishes on the surface of the volume of integration, it is found that the first term corresponds to the term $(\text{curl}\,(\boldsymbol{\xi}\wedge\mathbf{H}_0))^2$ in H^2, but that the second term does not take the required form. It is the total potential energy which matters, however, including other forms of energy such as the thermal energy of the gas, and Hain, Lüst and Schlüter (1957) have shown that δI_2 does take the required form, when the unperturbed system is in equilibrium. A general proof can indeed be given for any conservative system, the result being essentially Rayleigh's Reciprocal Theorem (see G. Temple and W. G. Bickley, *Rayleigh's Principle*, Oxford University Press, 1933, p. 52). Hain, Lüst and Schlüter (1957) and Bernstein, Frieman, Kruskal and Kulsrud (1958) have developed variational methods and have used them on problems of stability, which are discussed in the next section by less refined methods.

5.6. Criteria of stability

The study of stability is a very important part of the literature of hydromagnetics, and is generally attacked by the method of small perturbations, which is essentially the same as the method used for the study of oscillations. The stability of magnetostatic models has been discussed in §§ 3.7 and 4.4. Most of the other problems are concerned with models which would be unstable in the absence of a magnetic field (or if the material were not conducting) and it is often found that the field has a stabilizing effect, so that the model becomes stable when the field is strong enough.

In the method of small perturbations, the equations are linearized: each term is of first order in the amplitude of the perturbation. In the important examples which involve shear flow the first-order contributions of the inertial force $-\rho(\mathbf{u}\cdot\nabla)\,\mathbf{u}$ must not be forgotten. The perturbation is taken to vary with time as $e^{i\omega t}$, and the equations and boundary conditions determine the possible values of ω. A negative imaginary value of ω indicates that the perturbation increases in magnitude exponentially in time, so that the system will depart more and more from its unperturbed state, at least until the linear equations become invalid. The existence of a mode with a negative imaginary value of ω therefore means that the model is unstable. In the simpler problems ω is always pure imaginary for the unstable modes, and the critical condition then corresponds to $\omega = 0$. If the imaginary value of ω increases with increasing H, stability requires that H should exceed that value for which $\omega = 0$.

Often the value of ω can be complex for unstable modes; it is then said that 'the principle of the exchange of stabilities is invalid'. A value of ω with $\mathscr{R}(\omega) \neq 0$, $\mathscr{I}(\omega) < 0$ represents an oscillation, whose amplitude grows exponentially in time. This is sometimes known as 'overstability', but is a form of instability, since it leads to increasing departures from the unperturbed state. It may lead to the formation of fairly regular waves, as in the case of waves on water, or to a more turbulent state. The critical condition is generally $\mathscr{I}(\omega) = 0$, and the theoretical problem is more complicated than when the exchange of stabilities is valid. If the dissipative effects are neglected, however, ω is real for a range of conditions and the critical condition is the occurrence of a double root ω_1 of ω. Near the double root the equation can be approximated by

$$(\omega - \omega_1)^2 \approx \beta(H - H_c),$$

where β is positive, say. Then, if H is slightly less than H_c, the roots are a conjugate complex pair, representing one growing and one damped oscillation. A relatively simple model of this type will now be discussed as an illustration of the method.

Consider a uniform incompressible inviscid perfectly conducting fluid rotating with angular velocity $\Omega(r)$, r and z being cylindrical co-ordinates referred to the axis of rotation. Let the magnetic field be everywhere parallel to the axis and of magnitude $H(r)$. The perturbation is represented by a velocity $\mathbf{u}$, a field $\mathbf{h}$ and a perturbation ϕ in the total pressure, and all of these are taken to be proportional to $\exp[\mathrm{i}(\omega t + m\theta + kz)]$, where m is an integer and k is real. In cylindrical co-ordinates the condition of incompressibility is

$$\frac{\partial u_r}{\partial r} + \frac{u_r}{r} + \frac{\mathrm{i}mu_\theta}{r} + \mathrm{i}ku_z = 0. \tag{5.26}$$

It is convenient to write $\omega + m\Omega = \omega'(r)$ and the linear terms in (3.4) then give

$$\omega' h_r = kHu_r, \tag{5.27}$$

$$\omega' h_\theta = kHu_\theta - \mathrm{i} r h_r \,\mathrm{d}\Omega/\mathrm{d}r, \tag{5.28}$$

$$\omega' h_z = kHu_z + \mathrm{i} u_r \,\mathrm{d}H/\mathrm{d}r. \tag{5.29}$$

The linear terms in the equation of transfer of momentum give

$$\mu(\omega' u_r + 2\mathrm{i}\Omega u_\theta) = \mathrm{i}\,\partial\phi/\partial r + 4\pi kHh_r, \tag{5.30}$$

$$\mu(\omega' u_\theta - \mathrm{i} u_r(2\Omega + r\,\mathrm{d}\Omega/\mathrm{d}r)) = -m\phi/r + 4\pi kHh_\theta, \tag{5.31}$$

$$\mu\omega' u_z = -k\phi + 4\pi kHh_z - 4\pi \mathrm{i} h_r \,\mathrm{d}H/\mathrm{d}r. \tag{5.32}$$

Now it is convenient to write $\omega_A(r) = kV_A(r)$ and

$$\omega''(r) = \omega' - \omega_A^2/\omega'. \tag{5.33}$$

Then eliminating $\mathbf{h}$ from equations (5.27) to (5.32) gives

$$\mu(\omega'' u_r + 2\mathrm{i}\Omega u_\theta) = \mathrm{i}\,\partial\phi/\partial r, \tag{5.34}$$

$$\mu\left(\omega'' u_\theta - \mathrm{i} u_r\left(2\Omega + \frac{\omega''}{\omega'} r \frac{\mathrm{d}\Omega}{\mathrm{d}r}\right)\right) = -\frac{m\phi}{r}, \tag{5.35}$$

$$\mu\omega'' u_z = -k\phi. \tag{5.36}$$

Now in the particular case, when both Ω and H are independent of r, substitution of these equations into (5.26) gives

$$\frac{\partial^2\phi}{\partial r^2} + \frac{1}{r}\frac{\partial\phi}{\partial r} - \frac{m^2\phi}{r^2} + k^2\left(\left(\frac{2\Omega}{\omega''}\right)^2 - 1\right)\phi = 0, \tag{5.37}$$

whose solutions are Bessel functions. Consider then a model in which Ω and H are constant inside a cylinder $r < R$, and zero outside $r > R$, the change at $r = R$ being discontinuous. The discontinuity in the velocity tends to set up surface waves, and the centrifugal force also has a destabilizing effect; a similar model was discussed qualitatively in § 5.3.

The solution ϕ_1, for $r < R$, must be finite when $r = 0$ and is proportional to $J_m[((2\Omega/\omega'')^2 - 1)^{\frac{1}{2}} kr]$ and the solution ϕ_2 for $r > R$ must tend to zero as $r \to \infty$ and is proportional to $K_m(kr)$, the modified Bessel function of the second kind. The matching of ϕ_1 and ϕ_2 at the boundary $r = R$ determines the possible frequencies and is now discussed.

The radial displacement ξ_r of the boundary is $-iu_r/\omega'$ and this displacement causes a zero order change in u_θ (of amount $R\Omega$) in a region whose volume is proportional to ξ_r. This can be represented in the linear equations by considering u_θ to behave like a Dirac δ-function; $\mathrm{d}\Omega/\mathrm{d}r$ behaves like a δ-function at the boundary and (5.35) shows that u_θ behaves like $\xi_r r\,\mathrm{d}\Omega/\mathrm{d}r$. To be consistent u_r/ω' must be continuous at the boundary and this may be confirmed from (5.26) remembering that ω' is discontinuous. Now (5.34) determines the discontinuity in ϕ; it may be expressed by

$$\phi_2(R)-\phi_1(R)=\xi_r\mu R\Omega^2, \tag{5.38}$$

which is seen to express the effect of the centrifugal force acting over the distance ξ_r. Now let

$$q_1=\frac{R}{\phi_1(R)}\left(\frac{\partial\phi_1}{\partial r}\right)_{r=R}$$

Equations (5.34) and (5.35) can be used to express $\xi_r/\phi_1(R)$ as a function of q_1, and (5.38) then gives an expression for ϕ_2/ϕ_1 as a function of q_1. Defining q_2 in the same way an expression for ϕ_2/ϕ_1 as a function of q_2 is obtained, and by equating these expressions an equation is obtained to determine ω. This reduces to

$$\left(\frac{4\Omega^2-\omega''^2}{q_1\,\omega''+2m\Omega}+\frac{\omega''}{q_2}\right)\frac{\omega'}{\Omega^2}=1+\frac{m\omega''}{q_2\Omega}. \tag{5.39}$$

For a given value of ω'' this determines one value of ω' and then (5.33) determines one value of ω_A^2, which need not be positive, however. The condition of critical stability requires a double root for ω and this occurs when the value of ω_A^2 regarded as a function of ω has a turning point; it will only be physically significant, however, if the value of ω_A^2 at the turning point is positive. The existence of a minimum with ω_A^2 positive will now be demonstrated.

Let m and Ω be positive. From the properties of Bessel functions q_2 (which is independent of ω) is less than $-m$. When $|\,\omega''\,|>2\Omega$, q_1 is greater than m. It may be noted that, when $\omega''=-2\Omega$, $q_1=m$ and (5.39) is satisfied with any value of ω', but this does not affect the present discussion. As $|\,\omega''\,|$ increases, q_1 increases monotonically having infinities at the zeros of $J_m[((2\Omega/\omega'')^2-1)^{\frac{1}{2}}kR]$. Consider now how ω_A^2 behaves as ω'' increases from $-\infty$. When

$\omega''\to\pm\infty$, ω' tends to a finite positive limit (assuming $m\neq 0$) and hence, when $\omega''\to-\infty$, $\omega_A^2\to+\infty$. When $\omega''<-2\Omega$, $\omega'>0$ and hence $\omega_A^2>0$. As ω'' continues to increase ω_A^2 will have infinities when $\omega'\to\infty$ and zeros when $\omega'=0$ or $\omega'=\omega''$. These are clustered near $\omega''=0$, but we here consider the order in which they occur as ω'' increases from -2Ω.

When ω'' is slightly greater than -2Ω, $(q_1\omega''+2m\Omega)$ is positive and so long as it remains positive ω' is finite and positive and hence cannot equal ω'', so that there are no infinities or zeros of ω_A^2. Then q_1 passes through an infinity and $(q_1\omega''+2m\Omega)$ becomes large and negative. The expression in brackets on the left of (5.39) is still positive, but, as ω'' increases further, it decreases continually. Since zeros of ω_A^2 require either negative or infinite values of this bracket, and an infinity in ω_A^2 requires a zero value, an infinity occurs at a lower value of ω'' than any zeros. It then follows that ω_A^2 has a positive minimum at some still lower value of ω''. This is then a critical value for stability with respect to the mode specified by m and k.

The critical value may be illustrated by making approximations which are likely to be valid when kR is very small. Then q_2 is large and $q_1\approx m$, except when ω''/Ω is very small. Then (5.39) can be approximated by

$$(2\Omega-\omega'')\,\omega'=m\Omega^2,$$

which leads to

$$\omega_A^2=(\omega'-\Omega)^2+(m-1)\,\Omega^2.$$

The critical value occurs when $\omega''=(2-m)\,\Omega$, so that the approximations should be satisfactory for $m>2$. The critical value of H is proportional to $(m-1)^{\frac{1}{2}}$. This idealized model is therefore always unstable with respect to modes with a large enough value of m, but the dissipative effects are important for very high-order modes. They should therefore be included in studies of this kind, but an exhaustive treatment of any such problem would require too much space to be included here.

Some problems of stability have received a thorough treatment, many of them by Chandrasekhar and his collaborators; he has reviewed the subject in his George Darwin Lecture (1953). The classical problem of shear flow is the flow between rotating cylinders, which was first studied both experimentally and theoretically by

Sir G. I. Taylor. He assumed that the perturbation was axially symmetric. Chandrasekhar has studied the effect of a magnetic field on this system, when the fluid is conducting and finds that it has a stabilizing effect, which depends on the value of $H^2\sigma a^2/\mu\eta c^2$, where a is the distance between the cylinders. He found a similar, but less pronounced, effect on convectional stability. In the case of the gravitational instability of an infinite homogeneous fluid, however, Jeans's criterion is not affected by the field. He also worked out the effect of Coriolis forces together with a magnetic field on convectional and gravitational stability. In the former case the field has an important effect on the size of convection cells. Chandrasekhar and Fermi (1953*b*) discussed the gravitational stability of an infinite cylinder as a model for a spiral arm of a galaxy. In order to increase the time of instability from 10^8 years to $5 \cdot 10^9$ years they found that a field of $\sim 7 \cdot 10^{-6}$ gauss is required.

5·7. Magnetic variable stars

It is convenient to insert here a discussion of the magnetic variable stars discovered by H. W. Babcock. The best observed of these is HD 1252 which was known to be a spectrum variable with period about 9·3 days before Babcock, using the Zeeman effect, discovered that its magnetic field varies with the same period. From the sharpness of the spectral lines Babcock concluded that the axis of rotation is near the line of sight, and he measured the line of sight component of the magnetic field. This reverses in sign during the oscillations; using a dipole model for the field the polar field varies from $+7000$ to -6000 gauss.

The variations in the spectra of Babcock stars are complicated, and those of HD 125248 will only be outlined here. The spectral lines vary in brightness, but lines belonging to some elements are brightest at the positive maximum of the field, while others are brightest at its negative maximum. The radial velocities of different elements all show a similar anharmonic variation, but differing in magnitude. The largest velocities are a few kilometres per second; during an oscillation such a velocity would displace the material through a considerable fraction of a stellar radius. A notable feature of the velocity curve is a sudden acceleration downwards on the star, which occurs just before the positive maximum of the field.

The variation of the field is not sinusoidal; the positive maximum has a flat top while the negative maximum is roughly sinusoidal. When the field is changing sign between alternate maxima, a 'cross-over' effect occurs, which suggests that there are two distinct zones on the surface of the star with opposite polarity (in the line of sight) and slightly different velocity.

Since only the integrated light of a star can be observed, no detailed comparison with a model can be made, and the only possible contribution of theory is to find the simplest model which is consistent with the observations of the integrated light. Cowling (1952) has discussed the simplest possible model, and has succeeded in showing that its theoretical behaviour cannot be made to fit the observations by any reasonable means. It is assumed that the field moves with the material and the kinematical problem will first be discussed. The observed Zeeman effect measures the average line of sight component of the field, the average being weighted according to the amount of light received from each zone of the stellar disc; the way in which the intensity varies over the disc is roughly known. The observations require that this average field should change sign as a result of the oscillation. Cowling investigated the effect of a plausible form of distortion on a dipole field with its axis along the line of sight, and found that an unreasonably large distortion is required to make the equatorial polarity predominate over the polar polarity. It therefore seems desirable to consider the next simplest possibility, which is still axially symmetric, and is illustrated in fig. 5.4, where the axis of symmetry is vertical and the quadrant shown lies in the visible hemisphere of the star. The extreme phases of an oscillation are shown: in (*a*) the equatorial contribution to the line of sight field is small, and in (*b*) the polar contribution is small. This model therefore satisfies the kinematic requirement.

The period of the oscillations must now be discussed. The observations show that the motion is not perpendicular to the meridian planes (as it would be in torsional oscillations). Now a star without a magnetic field has modes of oscillations of this type, and Cowling showed that the periods of the few most fundamental modes are an order of magnitude less than the observed period of HD 125248 and pointed out that the magnetic field could only reduce the period. He next considers high-order modes, which he calls 'gravity oscillations'. In these the motion is nearly horizontal and there are a number of nodal surfaces, on which there is no vertical motion. The force due to gravity is then much diluted by having to drive the larger horizontal motion, and there are a large number of modes with periods of the right order of magnitude.

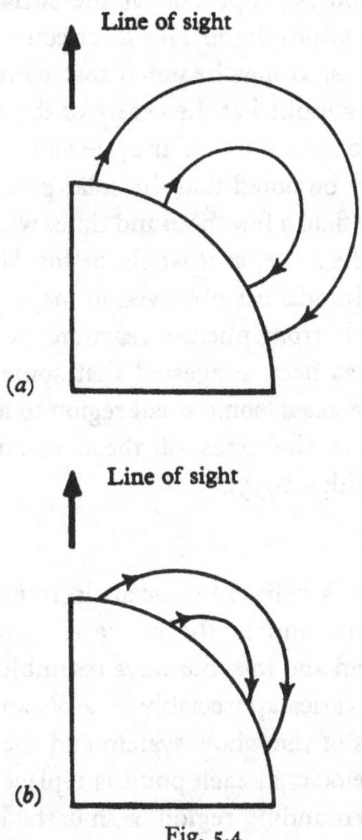

Fig. 5.4

Furthermore, the vertical gradient of the horizontal velocity is large, and so can cause large changes in the direction of the field at the surface and hence large changes in the line of sight component. The energy required to maintain the oscillations would probably be supplied in the outer layer of the star, presumably as a result of convective instability.

A further possibility arises from the fact that the amplitude of the oscillations is large, the displacement being comparable to the stellar radius. Non-linear effects may therefore be important. Torsional oscillations were rejected because they could not affect the line of sight component of the field in the linear approximation, but their non-linear effects may be important. A strong twisting would cause a force directed towards the axis and, since $H^2/8\pi \sim 10^6$ dynes/cm.^{2} which is larger than the gas pressure at the stellar surface, this force would be important at the surface and very important in the stellar atmosphere. This force could drive the motion in the meridian planes. It may be noted that with this mechanism the energy could be supplied at the centre of the star.

The relationship between the solar cycle and magnetic variables is intriguing. It may be noted that the solar period is nearly 1000 times longer and the field a few thousand times weaker, so that their product is roughly the same, as it would be for Alfvén waves.

The anomalous abundances observed in magnetic variable stars are believed to result from nuclear reactions occurring near the stellar surface. It has been suggested that some electromagnetic mechanism heats the gas in some small region to a sufficiently high temperature to make the rates of these reactions appreciable (Burbidge and Burbidge 1955).

5.8. Turbulence

Turbulent motion is believed to occur in interstellar space and in stellar atmospheres, and in the convective zones of stars the motion is complicated and in some ways resembles turbulence. In turbulent motion $\mathbf{u}$ varies appreciably in a distance much smaller than the dimensions of the whole system and the stream lines are 'jumbled'. If the velocity at each point is replaced by the average velocity of some surrounding region, a smoothed out picture will result; by sufficient smoothing a relatively simple average motion is obtained, which is known in the literature of turbulence as 'the large eddies'. Viscosity is important only for eddies whose scale is not much larger than η/u, where u refers to the motion as a whole. The scale of the large eddies exceeds this by the factor au/η which is the Reynolds number and must be large for turbulence to occur. The eddies of intermediate size are known as 'the inertial range'. Energy is fed from the large eddies to the dissipative eddies via the inertial eddies. The inertial eddies are treated statistically in the theory of turbulence, and their effect on the large eddies can be expressed as an effective increase in the coefficient of viscosity.

The effect of turbulence on a magnetic field is easily seen by considering the motion of the field; the primary effect is obtained by neglecting the motion of the field relative to the material. The small-scale motion jumbles the lines of force and this will stretch them, because the motion of two neighbouring points is similar to a random walk and on the average the distance between the two points increases. The stretching of the lines of force is opposed by the magnetic force and the nature of the resulting motion depends on the ratio of the energy densities of the undisturbed field $\mathbf{H}_0$ and of the turbulence. When $H_0^2 \gg 4\pi\mu u^2$ the turbulence cannot increase the length of the lines of force by an appreciable factor, and in this case the field inhibits turbulence. When $H_0^2 \sim 4\pi\mu u^2$ the form of the turbulence must still be strongly influenced by the field, but, when $H_0^2 \ll 4\pi\mu u^2$, it should be practically the same as if there were no field. In the latter case the lines of force become so jumbled that the field has 'small eddies' containing more energy than the 'large eddies' described by $\mathbf{H}_0$. The magnitude of the current density is increased, and hence also the Ohmic field, which can then be important, and will be discussed shortly.

Some aspects of the theory of turbulence are clarified by working in terms of the spatial Fourier transform $\mathbf{u}_\mathbf{k}$ of $\mathbf{u}$ and similar Fourier transforms of $\mathbf{H}$ and the other variables. The magnitude of $\mathbf{u}_\mathbf{k}$ as a function of $\mathbf{k}$ defines the 'spectrum of turbulence'; it should be noted that this refers to the wave number and not to the frequency. The large and small eddies are described by the small and large wave numbers respectively. When there is a magnetic field, the equations of motion in terms of the Fourier transforms are the equations for Alfvén waves modified by the inclusion of the non-linear terms, which couple different wave numbers. Thus turbulence in a magnetic field can be regarded as a superposition of Alfvén waves. It was previously concluded that turbulence is almost inhibited when $H_0^2 \gg 4\pi\mu u^2$; in this case $h^2 \sim 4\pi\mu u^2$ and the non-linear terms are relatively unimportant, so that the turbulence consists of weakly coupled Alfvén waves. When $H_0^2 \ll 4\pi\mu u^2$ the non-linear terms are the most important, as in the absence of a field.

When $H_0^2 \ll 4\pi\mu u^2$ and the field moves with the material, it is to be expected that the small eddies of the field will continue to receive energy from the turbulence, until this is prevented by the magnetic force. It was first suggested by Alfvén that this should occur when the field energy is comparable to that of the turbulence. The Ohmic field must be considered, however, because the decay of the small eddies of the field might be rapid. Batchelor's (1950) discussion of this problem will now be outlined. The rate of change of magnetic energy is given by

$$\tfrac{1}{2}\partial\overline{H^2}/\partial t = \overline{\mathbf{H}.\operatorname{curl}(\mathbf{u}\wedge\mathbf{H})} + c^2\overline{\mathbf{H}.\nabla^2\mathbf{H}}/4\pi\sigma, \qquad (5.40)$$

where the bars denote averages over a large volume. It is assumed that $\mathbf{H}_0$ is a negligible part of $\mathbf{H}$ and that $\mathbf{H}$ is a random function, such that the following average values all vanish: $\overline{H^2\nabla.\mathbf{u}}$, $\overline{\mathbf{u}.\nabla H^2}$, $\overline{\nabla^2 H^2}$ and $\overline{\nabla.(\mathbf{H}.\nabla)\mathbf{H}}$. Then (5.40) reduces to

$$\tfrac{1}{2}\partial\overline{H^2}/\partial t = \overline{\mathbf{H}.(\mathbf{H}.\nabla)\mathbf{u}} - c^2\overline{|\operatorname{curl}\mathbf{H}|^2}/4\pi\sigma. \qquad (5.41)$$

The first term on the right-hand side expresses the rate at which the lines of force would be stretched, if they moved with the material, and the second term is the rate of Joule dissipation.

It is now supposed that the magnetic force is unimportant and then the rate of change of the mean square vorticity $\overline{\boldsymbol{\omega}^2}$ is described by an equation of exactly the same form as (5.41); in it $\mathbf{H}$ is replaced by $\boldsymbol{\omega}$ and $c^2/4\pi\sigma$ by η. The mean square vorticity is increased by the stretching of the vortex lines and decreased by viscous dissipation.

Now in astrophysical systems there is usually a statistical steady state, so that these two effects are in equilibrium; then

$$\overline{\boldsymbol{\omega}.(\boldsymbol{\omega}.\nabla)\mathbf{u}} = \eta\overline{|\operatorname{curl}\boldsymbol{\omega}|^2}. \qquad (5.42)$$

This condition is also found to be approximately true in experiments at large Reynolds number.

At this stage Batchelor made the assumption that, since the field is controlled by the turbulence and since its equation of change has the same form as that for ω, H has the same statistical properties as ω. This assumption is heuristic, and has been criticized by Chandrasekhar (1955), but has yet to be replaced by a more satisfactory argument. According to Batchelor then

$$\overline{\mathbf{H}.(\mathbf{H}.\nabla)\mathbf{u}} \approx \eta\overline{|\operatorname{curl}\mathbf{H}|^2},$$

and then from (5.41) $\partial\overline{H^2}/\partial t \gtrless 0$ as $4\pi\sigma\eta/c^2 \gtrless 1$. It may be remembered that the relative importance of viscosity and the Ohmic field in the damping of Alfvén waves depends on the same criterion. Then, if $4\pi\sigma\eta/c^2 > 1$, $\overline{H^2}$ increases until the magnetic force is large enough to invalidate the argument and this occurs when $\overline{H^2} \sim 4\pi\mu\overline{u^2}$. It is important to note that this does not depend on $\mathbf{H}_0$. If $4\pi\sigma\eta/c^2 < 1$, $\overline{H^2}$ would decrease, if $\mathbf{H}$ had the same statistical properties as $\boldsymbol{\omega}$, and hence $\overline{H^2}$ cannot greatly exceed H_0^2.

The rate of change of the mean field $\mathbf{H}_0$ due to the Ohmic electric field is increased by turbulent motion. This was demonstrated by Sweet (1950) for the axially symmetric case and will here be explained for the case of two-dimensional symmetry. The concept of the motion of the field relative to the material (see § 3.3) will be used; because of the symmetry $V=0$ and the velocity of the field relative to the material is $c\mathbf{j}\wedge\mathbf{H}/H^2\sigma$.

Let all quantities be independent of z and let the vector potential be $(0, 0, A)$. The effect of the turbulence is to distort the surfaces of constant A, which are taken to move with the velocity of the field. A measure of the effect of the Ohmic field is given by the rate of change of the quantity of material enclosed by such a surface per unit length in the z-direction. This is negative and with an appropriate convention for the sign of j is M given by

$$M = c\sigma^{-1}\oint \mu j H^{-1}\,\mathrm{d}s, \tag{5.43}$$

the integral being taken round a line of force on the surface. Now H is $\delta A/\delta n$, where δn is the distance between the lines of force defined by the values A and $A+\delta A$, so that (5.43) is equivalent to

$$M = c(\sigma\delta A)^{-1}\int \mu j\,\mathrm{d}S,$$

where the integral is taken over the region between these two lines of force.

Putting $4\pi\mathbf{j} = c\,\mathrm{curl}\,\mathbf{H}$ and transforming by Stokes's theorem then gives

$$M = c^2(4\pi\sigma\delta A)^{-1}\left\{\oint_{A+\delta A}\mu H\,\mathrm{d}s - \oint_{A}\mu H\,\mathrm{d}s + \oint_{\delta A}\mathbf{H}\wedge\nabla\mu\,\mathrm{d}S\right\}. \tag{5.44}$$

It will be assumed that $\nabla\bar{\mu}=0$, so that the average value of the last term vanishes and then the effect of turbulence is made clear by (5.44). In the primary effect of turbulence the field moves with the material, so that for any short length of a line of force $H \propto \mu\,ds$. The stretching of the lines of force therefore increases $\oint \mu H\,ds$ by a factor whose average value is $\overline{H^2}/H_0^2$; it therefore increases M by this factor and this can be expressed as an effective reduction of the conductivity.

A similar effect is expected to occur in the general case, but this has not been treated mathematically. If the motion of the field is used, a cut must be made as in § 3.3.

5.9. Shock waves

Macroscopic motion with supersonic speeds sometimes occurs in stellar atmospheres, and then shock waves are likely to be generated. The influence of a magnetic field on a shock wave in a conducting fluid has been studied by several authors. If **H** is perpendicular to the shock front, the field has no effect, and here **H** will be taken to be parallel to the shock front. The conductivity will be taken to be infinite; Marshall (1955) has worked out models for various values of the conductivity and including the Hall effect. We work in the frame moving with the shock, so that there is a steady state, and take the x-axis perpendicular to the shock front. The velocity is u in the x-direction. The conservation of matter requires μu to be constant through the flow and Hu is also constant, because the field moves with the material. Using (2.17) the equation of transport of momentum is

$$\mu u \frac{du}{dx} = -\frac{dp}{dx} - (8\pi)^{-1}\frac{dH^2}{dx} + \frac{4}{3}\frac{d}{dx}\mu\eta\frac{du}{dx}$$

and can be integrated directly. It is convenient to write the result as

$$\tfrac{4}{3}\mu\eta\frac{du}{dx} = nk(T-\theta(u)), \tag{5.45}$$

where $$nk\theta = n_0kT_0 + \mu u(u_0-u) + (Hu)^2(u_0^{-2}-u^{-2})/8\pi, \tag{5.46}$$

and the subscript zero refers to the limiting values at a great distance in front of the shock; in the limit conditions are uniform, so that du/dx must vanish, and this is ensured by (5.46).

The variation of temperature is determined by (2.10) which becomes

$$\tfrac{3}{2}nku\frac{dT}{dx}=\frac{d}{dx}\left(K\frac{dT}{dx}\right)-\left(p-\tfrac{4}{3}\mu\eta\frac{du}{dx}\right)\frac{du}{dx}, \tag{5.47}$$

where K is the thermal conductivity, and heating by collisions between unlike particles is neglected, because it corresponds, in the case of a totally ionized gas, to Joule heating and is therefore inversely proportional to the conductivity. Using (5.45), (5.47) can be integrated to give

$$K\frac{dT}{dx}=nuk\left(\tfrac{3}{2}(T-T_0)+\int_{u_0}^{u}(\theta/u)\,du\right), \tag{5.48}$$

the constant of integration having been chosen so that dT/dx vanishes for the limiting values u_0 and T_0.

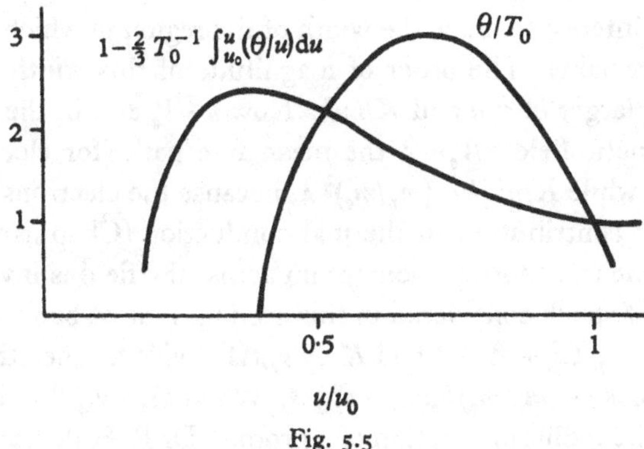

Fig. 5.5

Since n must remain positive, u cannot change sign; taking u_0 to be positive, only positive values of u can occur. Fig. 5.5 shows an illustration of θ plotted against u and also $T_0-\frac{2}{3}\int_{u_0}^{u}(\theta/u)\,du$, which is the value of T for which dT/dx vanishes for a given value of u. These curves must intersect at (u_0, T_0). The condition for them to touch is $d\theta/du=-2\theta/3u$ and it is found from (5.46) that this occurs when u_0 is the modified velocity of sound V_s' given by (5.21). For a compression shock u_0 must exceed this value and this will now be assumed; rarefaction shocks are known to be unstable (Sen, 1956). The curves then intersect again at (u_1, T_1) with $0<u_1<u_0$ and $T_1>T_0$. As u and T tend to these values, uniform conditions are again approached, and so they can be interpreted as the limiting values at a large distance behind the shock front. The Rankine–Hugoniot relations between u_1, T_1, u_0 and T_0 are determined by the equations

$$T_1 = \theta_1 = T_0 - \frac{2}{3}\int_{u_0}^{u_1} (\theta/u)\,du. \tag{5.49}$$

Substituting from (5.46) into (5.49) gives, after considerable rearrangement,

$$4\frac{u_1}{u_0} = 1 + 3\left(\frac{V_s'}{u_0}\right)^2 + \frac{1}{2}\left(\frac{u_0}{u_1} - 1\right)\left(\frac{V_A}{u_0}\right)^2, \tag{5.50}$$

where V_s' and V_A refer to the initial values. Since $u_0 > u_1$ (5.50) shows that the presence of a magnetic field reduces the compression ratio u_0/u_1 more than it would be reduced by increasing T_0 to give the same velocity of sound.

It is of interest to know the width of the region in which u and T vary appreciably. The order of magnitude of this width is a few times the larger of η/u and K/nuk. Now $u \sim V_s$ and in the absence of a magnetic field $\eta/V_s \sim \lambda$ the mean free path (for electrons or protons), while $K/nV_s k \sim (m_p/m_e)^{\frac{1}{2}} \lambda$, because the electrons provide the major contribution to thermal conduction (Chapman, 1954). Because the transport of momentum across the field is involved in the type of shock considered in this section, η must be reduced by the factor $\nu_{pp}(\Omega_p^2 + \nu_{pp}^2)^{-\frac{1}{2}}$ and K by $\nu_{ee}(\Omega_e^2 + \nu_{ee}^2)^{-\frac{1}{2}}$. Then the width of the shock $\sim (m_p/m_e)\, V_s(\Omega_e^2 + \nu_{ee}^2)^{-\frac{1}{2}}$. When $\Omega_e > \nu_{ee}$ this is of the order of the radius of gyration of a proton. Dr R. Hide has pointed out to the author, however, that the above treatment is not valid when $\Omega_e > \nu_{ee}$, because the distribution function behind the shock is then non-thermal.

CHAPTER 6

ACCELERATION OF CHARGED PARTICLES TO HIGH ENERGY

6.1. Requirements of an accelerating mechanism

All primary cosmic rays have energies greater than 10^9 eV. and it is believed that some have energies as high as 10^{17} eV. The mechanism by which they acquire their energy is presumed to be acceleration by some electric field, because no other plausible mechanism is known by which such large energies can be achieved. The origin of cosmic rays is therefore a problem of cosmic electrodynamics. An electrical accelerating mechanism is probably also required for the primary auroral particles, whose energy is ~ 1 MeV.

It has been seen that the principal contribution to the electric field is the induced field $-\mathbf{u}\wedge\mathbf{H}/c$. The order of magnitude of the voltage is given by $10^{-8}\, auH$ volts; with values taken from table 1.1 this is 10^9 volts for the chromosphere above sunspots, 10^{10} volts for the corona and 10^4 volts for H II regions. It will be seen later in this chapter that the induced field can accelerate particles to very high energies, but only in special circumstances. The reasons why the electric field does not generally accelerate the particles are discussed here. In nearly uniform fields, for instance, since $\mathbf{E}$ is perpendicular to $\mathbf{H}$ and $E<H$, the particles just drift with the velocity $\mathbf{u}$, and their energies vary only by small amounts. Again for non-uniform fields but uniform velocity $\mathbf{u}$, $\mathbf{E}$ can be made to vanish by a Lorentz transformation, and then in the new frame the energy of a particle is constant; then, since $u \ll c$, large energies cannot be achieved in the original frame. It follows that the existence of the induced field is by no means sufficient to obtain an accelerating mechanism. A field is required in which a particle's orbit is such that the electric field does a large amount of work on the particle.

Before discussing orbits a general consideration of energy is useful. The particles acquire their energy directly from the electric field and this draws energy from the magnetic field, which in turn can be replenished from the kinetic energy of bulk motion in the gas. Suppose now that the field energy and kinetic energy in some region of space is released and distributed *equally* among the charged particles in the same region; the resulting acceleration of the particles may be regarded as an increase in temperature and, unless the density of field or kinetic energy is much larger than the initial gas pressure, the temperature is not increased by a large factor. Such a distribution of energy does not give the particles anything like the energy sought in this chapter. A mechanism is required which somehow selects a small minority of the particles to receive the greater share of the field energy.

6.2. The betatron mechanism

The first mechanism suggested (Swann, 1933) was that particles are accelerated near sunspots by the same mechanism as that used in the betatron. It is supposed that the magnetic field may be taken to be uniform and that its magnitude increases with time, so that the electric field lines are circles all centred on an axis of symmetry, which is parallel to the magnetic field. The direction of the electric field is such as to drive a current opposing the change of magnetic field, and the charged particles spiral in directions such as to reduce the magnetic field inside the spiral; the particles are therefore accelerated by the electric field when H is increasing and vice versa. The energy gained by a particle is $e\int \mathbf{E} . d\mathbf{s}$ taken along its orbit.

The change in $\mathbf{H}$ occurring in one gyro-period is usually very small for cosmic systems (the gyro-period for protons is $10^{-4}H^{-1}$ sec.) and the spatial variation is usually small in the distance travelled by a particle in this time. Provided this is so, the temporal and spatial variations of $\mathbf{H}$ may be treated as perturbations for the purpose of calculating the orbits of charged particles (Alfvén, 1950*a*). The electric field may also be treated as a perturbation. The unperturbed orbit corresponds to a uniform magnetic field and is helical (see § 1.5); denoting the components of the particle's velocity parallel and perpendicular to $\mathbf{H}$ by $v_{\parallel}$ and $v_{\perp}$, the radius of the cylinder on which the helix lies is $r = mcv_{\perp}/eH$. The changes in $v_{\parallel}$ and $v_{\perp}$ are obtained by integrating the perturbation over one turn of the helix. The change in energy depends on $\int \mathbf{E}\,.\,d\mathbf{s}$ and $\mathbf{E}$ is perpendicular to $\mathbf{H}$. Then Stokes's theorem gives for the first approximation

$$\int \mathbf{E}\,.\,d\mathbf{s} = \pm \pi r^2 c^{-1} \partial H/\partial t,$$

the sign depending on the sign of the charge of the particle. Then, since r is $mcv_\perp/eH$, the average rate of acceleration is given by

$$\frac{d(mv_\perp)}{dt} = \frac{mv_\perp}{2H}\frac{\partial H}{\partial t},$$

and then $$mv_\perp \propto H^{\frac{1}{2}}. \tag{6.1}$$

When special relativity is taken into account this is still valid if m is the relativistic mass.

When the perturbation method is valid, the particles must drift with the usual velocity given by (1.1) as well as gaining energy by the betatron mechanism. This means that the field is still frozen into the material, so that in order to increase H a compression is needed and the density of charged particles increases proportionally to H. This result depends on the assumption that the electric field is produced by induction; no other source of the electric field has been suggested. The existence of the electric field is usually inferred from the observed change of the magnetic field. The model in which the magnetic field is uniform is then misleading, however, since a change in H can result from the uniform motion of a non-uniform field, as explained in § 6.1.

When the perturbation method is valid, $(mv_\perp)^2$ and n are both proportional to H. This result is just the same as that for the adiabatic compression of a gas. The process is not peculiarly electromagnetic, but the magnetic field traps the particles and so takes the role of a box. The process is not plausible as an accelerating mechanism for the following reasons. There is no obvious way in which a few particles are selected for acceleration. In order to obtain high energy particles a narrow tube of force would have to be compressed by pressure from outside, and absorb energy from a much larger region. The magnetic and gas pressures in this narrow tube would need to increase by a very large factor, and there is no reason why this should happen. The process may be important as a heating mechanism, however: if the gas is alternately compressed and expanded at intervals of the order of the collision time, a considerable fraction of the energy of compression is turned into heat.

It remains to consider the possibility that acceleration occurs when the perturbation method is not valid, presumably in very weak fields. Then $c\,\mathrm{curl}\,\mathbf{E}$ is not much smaller than ΩH, where Ω is the gyro-frequency of the particles to be accelerated. Now, assuming $\mathbf{E}$ to be an induction field $E \sim uH/c$, and it follows that the region of uniformly increasing H cannot extend over a distance much greater than u/Ω; since this is the radius of the orbit of a particle of velocity u, it seems that particles with velocities much greater than u cannot be obtained by this mechanism. Certainly the model with a uniform field cannot generate high-energy particles, and, though this model can be modified in several different ways, none is known to show any promise of overcoming the difficulties which have appeared in this discussion.

6.3. Discharges in a magnetic field

In the simplest accelerating mechanism the electric field is roughly uniform and the particles move parallel to this field; this may be called a 'discharge'. If particles are accelerated by such a field, it cannot satisfy (1.1), for the particles would then just drift with velocity u. Since the positive and negative particles are accelerated in opposite directions in the accelerating region, however, they produce a large current density, and then (1.1) is invalid. Outside the accelerating region (1.1) is valid and the accelerating field may be regarded as being generated by a 'dynamo' mechanism. The first requirement for a discharge, then, is a large current density, or, since (1.3) is still valid, a large value of curl $\mathbf{H}$. Supposing the particles to be accelerated to 1 MeV or more, the electron velocity approaches the velocity of light and $j \sim nec$; putting $|\mathrm{curl}\,\mathbf{H}| \sim H/b$, then $b \sim H/4\pi ne$. The numerical values given in table 1 give $b = 1$ mm. for sunspots and 10 m. for both the Corona and interstellar space. Thus curl $\mathbf{H}$ must be vastly greater than the value given by the usual order of magnitude H/a, where a is a characteristic length of the system; as pointed out in § 1.3, such a large value can only occur in a thin region of space, the thickness in one direction being of order b. Although it is hard to imagine such a very thin region being important, it will be seen that the number of particles accelerated is appreciable. The thinness of the accelerating region is useful in selecting a few particles to receive energy.

The next important question is how curl **H** can be built up to a very large value, if it has its usual order of magnitude initially. Although (1.1) is invalid for a discharge, it is valid before the current density becomes too large, so that curl **H** must be increased by motion of the material. The motion is of course strongly in-

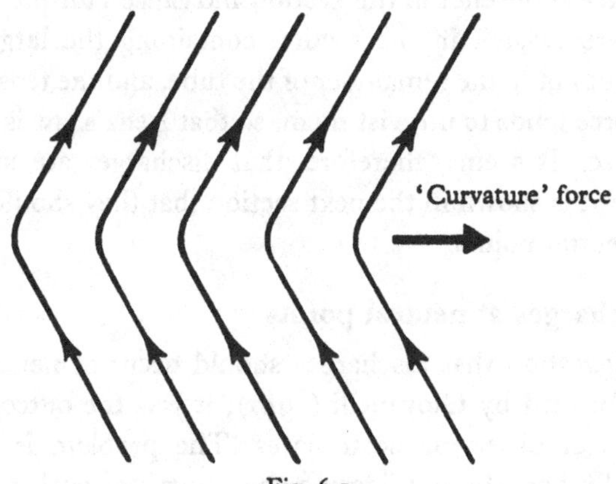

Fig. 6.1

fluenced by the magnetic force, particularly when the current density becomes large, and, since Lenz's law states that this force always opposes the motion, there appears to be an insuperable difficulty. Lenz's law, however, was invented for rigid conductors moving in a uniform magnetic field, and it is necessary to verify that it can be generalized to hydromagnetics; an exception will in fact be described in § 6.4.

In chapter 4, static models were considered with discontinuities in **H**, where curl **H** would be infinite, but it was pointed out in § 4.2 that such discontinuities should be replaced by transition layers in which the current density is not excessively large. If we now consider the sort of motion required to increase curl **H** in such a layer, it is seen to be opposed by the magnetic pressure. Similarly, if the field is distorted by a current sheet, which is perpendicular to the field, the lines of force being shown in fig. 6.1, the 'curvature' force opposes the distortion. For these two examples, then, Lenz's law is valid. Large current densities are generated in shock fronts in a magnetic field, but, from § 5.9, the width of a shock is at least $4\pi nm_p V_s c/H^2 \times$ the value of b required here.

If curl **H** is large and parallel to **H** the lines of force are twisted. If the current runs right round a tube of force, the lines of force are linked many times and, as seen in § 3.2, such linking cannot result from the motion of the material. If the large current density is confined to a section of a tube of force, the current must cross the lines of force at the ends of this section and cause a torque. The lines of force are twisted in the section containing the large current density, but not in the remainder of the tube, and the tension in the lines of force tends to untwist them, so that Lenz's law is also valid in this case. It seems, therefore, that discharges are unlikely to occur, but it is shown in the next section that they should occur at X-type neutral points.

6.4. Discharges at neutral points

The suggestion that discharges should occur at neutral points was put forward by Giovanelli (1947); it was the outcome of his experience of observing solar flares. The problem is discussed theoretically here, by considering the growth of curl **H** to a very large value as in the last section. Consider the field in the neighbourhood of an X-type neutral point; the field may be expressed as a Taylor expansion with the neutral point as origin and the leading term is $\partial H_i/\partial x_j \delta x_j$, where δx_i is the position vector. The three limiting lines of force have the directions of the principal axes of the tensor $\partial H_i/\partial x_j$, and the antisymmetrical part of this is related to curl **H**. If curl **H** is zero, $\partial H_i/\partial x_j$ is symmetrical and a well-known theorem states that the principal axes are orthogonal. If these are taken as Cartesian axes, $\partial H_i/\partial x_j$ is diagonal, and, since $\operatorname{div}\mathbf{H}=0$, the diagonal components cannot all have the same sign. If $\partial H_1/\partial x_1$ and $\partial H_2/\partial x_2$ have opposite signs and a current is then started in the x_3-direction, the direction of the field at points in the (x_1, x_2) plane lies in the (x_1, x_2) plane and the lines of force in this plane are as shown in fig. 6.2, the directions of the limiting lines of force being no longer perpendicular. The direction of the current is such as to produce a field directed clockwise and the magnetic force density is therefore directed as shown by the heavy arrows in fig. 6.2. The resulting flow stretches the material in the vertical direction and compresses it in the horizontal direction. Then, since the field moves with the material, the limiting lines of force rotate towards each other and the angle between them becomes smaller. This suggests that the current density is increased by the motion and that Lenz's law is incorrect in this case. This result is next confirmed mathematically with the simplification that the pressure gradient is omitted. The pressure gradient is not relevant to Lenz's law and is never likely to stop a flow of the sort shown in fig. 6.2; the result obtained in § 4.3 (ii) provides further justification for the belief that the flow is not stopped in this particular case.

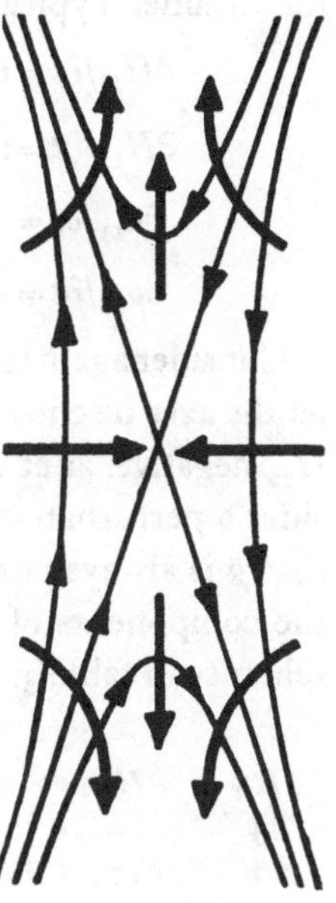

Fig. 6.2

In the frame in which the neutral point is initially at rest, the equations of motion with the pressure gradient omitted show that it remains a neutral point and stagnation point for all time. Writing u_{ij} and H_{ij} for the values of $\partial u_i/\partial x_j$ and $\partial H_i/\partial x_j$ at the neutral point, and operating with $\partial/\partial x_j$ on the equations determining $\partial H_i/\partial t$ and $\partial u_i/\partial t$ at the neutral point gives

$$\partial H_{ij}/\partial t = -u_{kj}H_{ik} + H_{kj}u_{ik} - H_{ij}u_{kk} \tag{6.2}$$

and

$$\partial u_{ij}/\partial t = -u_{kj}u_{ik} + (H_{il} - H_{li})H_{lj}/4\pi\mu. \tag{6.3}$$

These equations together with

$$\partial \mu/\partial t = -\mu u_{kk} \tag{6.4}$$

determine the time derivatives of u_{ij}, H_{ij} and μ in terms of these variables themselves. It is a rare stroke of luck to obtain a finite set of equations referring to a single point in a continuous system; if the point were not a neutral point or not a stagnation point, or if the pressure gradient had been included, higher derivatives would have been involved in (6.2) or (6.3). Apart from the pressure gradient, however, these equations contain the information required for the present problem.

To demonstrate the reversal of Lenz's law the problem may be simplified by taking all components of both H_{ij} and u_{ij} with either suffix equal to 3, except H_{33}, to vanish initially; they then vanish for all time. Typical equations for the other components are

$$\left.\begin{aligned} \partial H_{11}/\partial t &= u_{12}H_{21} - u_{21}H_{12} - H_{11}(u_{11}+u_{22}), \\ \partial H_{12}/\partial t &= u_{12}(H_{22}-H_{11}) - 2u_{22}H_{12}, \\ \partial u_{11}/\partial t &= -u_{1}^{2} - u_{12}u_{21} + (H_{12}-H_{21})\,H_{21}/4\pi\mu, \\ \partial u_{12}/\partial t &= -u_{12}(u_{11}+u_{22}) + (H_{12}-H_{21})\,H_{22}/4\pi\mu. \end{aligned}\right\} \quad (6.5)$$

Consider again the state in which the current density vanishes and let the axes be chosen so that H_{12} and H_{21} vanish, H_{11} is positive and H_{22} negative. Let all components of u_{ij} vanish initially and consider a perturbation in one of H_{12}, H_{21}, u_{12} and u_{21}. Remembering that μ is always positive, the equations (6.5) show that the signs of the components of H_{ij} and u_{ij} will at first be given by one of the schemes in table 3.

TABLE 3

H_{11}	H_{22}	H_{12}	H_{21}	u_{11}	u_{22}	u_{12}	u_{21}
+	−	−	+	−	−	+	+
+	−	+	−	−	−	−	−

It can then be seen that every term in the derivative of each component has the same sign as that component. Consequently all the components grow in magnitude, and, since H_{12} and H_{21} have opposite signs, the current density also grows. According to the equations (6.2) and (6.3) all the components continue to grow and become infinite in a finite time, because the right-hand sides of the equations are bilinear in the components. In practice of course one of the approximations made in (6.2) or (6.3) must become invalid. It was seen in § 4.3 (ii) that for a model with two-dimensional symmetry the condition of static equilibrium required an infinite current density at the neutral point, so that it is unlikely that the catastrophe is prevented by the pressure gradient. On the other hand, if no other approximation breaks down, the resistivity of the material will certainly limit the current density, though the limiting value is very large; a discharge then occurs and the simple theory of the conductivity is invalid.

The effect of the resistivity is easily seen by returning to the picture of the motion of the field and allowing for it to move relative to the material in the way explained in § 3.3. Here the lines of force may be restricted to lie on a plane, and no question of linking is involved. The Ohmic field causes a motion of the field relative to the material, the velocity $c\mathbf{j} \wedge \mathbf{H}/\sigma H^2$ having a singularity at the neutral point, and being important only in the region of large current density. The effect of this motion is to change two lines of force such as those in fig. 6.3 (*a*) into another pair such as those in fig. 6.3 (*b*) a short time later.

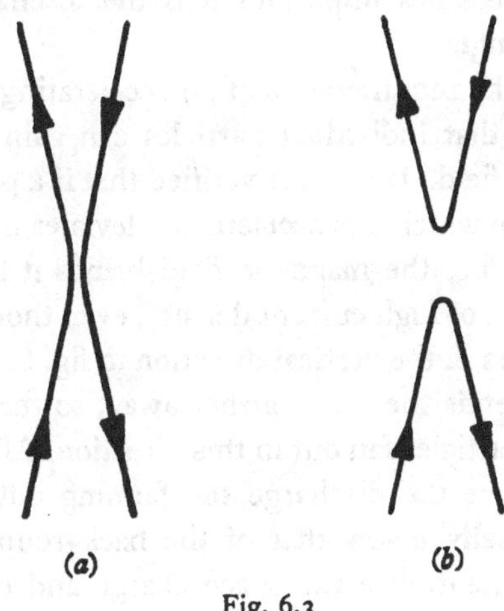

Fig. 6.3

The lines of force of fig. 6.3 (*a*) appear to have been broken and rejoined to form those of fig. 6.3 (*b*); Faraday's law shows that the rate at which lines of force undergo this process is equal to the electric field at the neutral point. The effect of the rearrangement of the field is to reduce the total length of the lines of force and hence the field energy. It also reduces the current density at the neutral point, and, when this rate of reduction balances the growth of current density due to the motion of the material, the field in the neighbourhood of the neutral point does not vary rapidly with time. The electric field is directed into the paper at all points in fig. 6.2 and under quasisteady conditions must be approximately uniform. Then the electric field at the neutral point is approximately the same as the induced field at points outside the discharge; it will be estimated for solar flares in § 7.8 and the voltage is found to be large.

Since the neutral point is destroyed by this mechanism it is difficult to understand how an X-type neutral point can occur at all. The time required for the growth of the current must be considered, however; this depends on the magnitude of the components of H_{ij}, and has the order of magnitude $(4\pi\mu)^{\frac{1}{2}} H_{ij}^{-1}$. A neutral point is probably formed between two separate regions of strong field and H_{ij} can be extremely weak. When H_{ij} is increased by the motion of the material sufficiently to make $(4\pi\mu)^{\frac{1}{2}} H_{ij}^{-1}$ comparable to the time scale of the normal changes of the system, the instability becomes important and the discharge cannot be delayed much longer.

To complete the requirements of an accelerating mechanism, it must be shown that individual particles can gain large energies from the electric field. It is easily verified that if a particle, moving in the direction in which it is accelerated, deviates in the horizontal direction in fig. 6.2, the magnetic field brings it back, so that it stays in the region of high current density, even though this is very thin. If it deviates in the vertical direction in fig. 6.2, however, the magnetic field bends the orbit farther away, so that the orbits of the accelerated particles fan out in this direction. After the accelerated particles leave the discharge the fanning out reduces their density substantially below that of the background ionized gas, which can then neutralize the space charge and current density of the accelerated particles.

6.5. Acceleration resulting from turbulence

Fermi's (1954) theory of the origin of cosmic rays supposes that there is a magnetic field in the interstellar gas in statistical equilibrium with turbulent motion there. The mechanism requires that the scale length for variations of the field be larger than the radius of the orbits of the particles accelerated, so that the orbits can be treated by the perturbation method. One further result obtained by this method is required for a description of Fermi's theory.

Let a tangent to the line of force on which the orbit is centred be the z-axis of cylindrical co-ordinates, and consider the orbit in the neighbourhood of the point of contact of this tangent. If $\overline{H_r}$ is the average value of H_r over a circle of constant r and z, (3.1) requires

$$\frac{1}{r}\frac{\partial}{\partial r}(r\overline{H}_r) = -\frac{\partial H_z}{\partial z}$$

assuming $\partial H_z/\partial z$ varies only slightly in distance r. Hence

$$\overline{H}_r = -\tfrac{1}{2}r\,\partial H_z/\partial z.$$

Now using the relativistic mass, $r = mv_\perp c/eH$ on the orbit, and

$$\frac{\mathrm{d}mv_\perp}{\mathrm{d}t} = -\frac{e\overline{H}_r v_\parallel}{c} = \frac{mv_\perp v_\parallel}{2H_z}\frac{\partial H_z}{\partial z} = \frac{mv_\perp}{2H_z}\frac{\mathrm{d}H_z}{\mathrm{d}t}.$$

Hence $mv_\perp \propto H^{\frac{1}{2}}$, which is the same as (6.1), and is therefore true for any variation such that the perturbation method is valid. If there is no betatron acceleration, the kinetic energy of the particle is constant. Then, if θ is the angle between $\mathbf{v}$ and $\mathbf{H}$, $\sin\theta$ $(=v_\perp/v)$ is proportional to $H^{\frac{1}{2}}$. Now there may be regions in which the field is so strong that this result would make $\sin\theta$ greater than 1; the particle cannot penetrate into such a region, and is turned back at the point where $v_\parallel = 0$ and $\sin\theta = 1$.

Fermi's mechanism involves 'collisions with the field' of which type A is the reflexion by a region of strong field just described. In type B the particle follows a line of force round a bend, and for simplicity it is assumed that its mean direction of motion is just reversed (the line of force turning through 180°). A charged particle in a turbulent field is continually suffering deflexions in the direction of its mean motion and slight energy changes, but Fermi simplified the problem by just considering 'collisions'; this simplification is somewhat similar to that made in describing the Coulomb interaction as equivalent to collisions in the Boltzmann theory of chapter 2. If the magnetic field were not moving, the energy of the particle would not be changed by a collision, but in a given frame of reference the field is in general moving. Fermi supposes that the motion may be taken as uniform over that region of the field, which is involved in the collision; owing to the non-uniformity of the field in such a region, the velocity of the field parallel to $\mathbf{H}$ then has a meaning. If the components parallel to $\mathbf{H}$ of the velocities of the particles and the field before a collision have opposite signs, it is a 'head-on' collision and the particle gains energy; if they have the same sign, it is an 'overtaking' collision and the particle loses energy.

The change of energy at a collision is calculated very simply by means of Lorentz transformations, and the detailed mechanism need not be known. In the given frame let W be the particle's energy including the rest energy, $\beta c = |\mathbf{v}|$ and θ be the angle between $\mathbf{v}$ and $\mathbf{H}$. Let Bc be the component of the velocity of the field parallel to $\mathbf{H}$ and be positive for a head-on collision. Then in the frame moving with the velocity Bc relative to the given frame the particle has energy $W(1+B\beta\cos\theta)(1-B^2)^{-\frac{1}{2}}$ and velocity component $(\beta\cos\theta+B)/(1+\beta B\cos\theta)$ parallel to $\mathbf{H}$. In this frame the collision leaves its energy unchanged and reverses its velocity. Transforming back to the given frame the energy W' after the collision is

$$W' = W(1+2B\beta\cos\theta+B^2)/(1-B^2),$$

which gives

$$\log(W'/W) = 2B(\beta\cos\theta+B-B\beta^2\cos^2\theta)+O(B^3). \qquad (6.6)$$

The accelerating mechanism depends partly on the fact that head-on collisions are more numerous than overtaking collisions. Their relative frequencies are proportional to the velocity of the particle relative to the field and so are in the ratio $(\beta\cos\theta+B)/(\beta\cos\theta-B)$. Then from (6.6) the average increase of $\log W$ per collision is

$$\overline{\delta\log W} = 2B^2(2-\beta^2\cos^2\theta)+O(B^3). \qquad (6.7)$$

Consequently, the particle gains energy from the turbulent field and W increases by a few powers of e after B^{-2} collisions. Fermi explained the principle of this accelerating mechanism in a simple way; he supposed that the accelerated particles have no collisions with other particles of the gas. The particles collide with lumps of the gas, whose scale is determined by the turbulence, so that the number of particles in each lump is enormous. The effect of the collisions is to attempt to establish equipartition of energy between the lumps and the accelerated particles, but the energy of the lumps is so large that an impossibly long time is required to establish equipartition. Thus the particles have much less energy than the lumps. The rate of acceleration then depends on the velocity of the lumps but not their mass, as is seen in (6.7).

Other processes, by which the accelerated particles lose energy, must now be taken into account. For electrons the loss of energy by radiation is important; with the numerical values adopted by Fermi the radiation loss exceeds the rate of gain of energy at all energies, so that electrons cannot be accelerated by this mechanism, and we now consider nuclei only. For nuclei the radiation loss is much less important, because of their greater mass, and, with Fermi's values, it is negligible. Ionization loss exceeds the rate of gain at energies below a certain threshold, but becomes negligible for energies well above the threshold; Fermi's estimate of the threshold was a fraction of 1 BeV. for protons and substantially higher values for heavier nuclei: his estimate was 300 BeV. for iron. These estimates depend on the rate of acceleration by the collisions with the field and on the density of the interstellar gas, and can therefore only be rough. It should be noted that this mechanism selects a small minority of the particles to be accelerated, namely those whose energies are already above the threshold.

Nuclear collisions have a somewhat different effect. From the absorption of primary cosmic radiation, in the high atmosphere, Fermi obtained a cross-section of about $2{\cdot}5 . 10^{-26}$ cm.^{2}, and, assuming a mean density in interstellar space of 10^{-24} g/cm.^{3}, the mean free time between nuclear collisions for a particle moving with the velocity of light is then about 60 million years. In such a collision a large amount of energy is probably converted into a spray of mesons, and the high energy nucleus making the collisions loses most of its energy; Fermi therefore regarded nuclear collisions as removing cosmic ray particles, but suggested that, when a nucleus well above the threshold energy makes a nuclear collision, several protons may be left with energies above their threshold; this would lead to a chain reaction, but a mechanism for the injection of heavy nuclei would still be needed. The removal of cosmic ray particles by nuclear collisions after an average time T ($\sim 6 . 10^7$ years) determines the form of the energy spectrum as follows.

Suppose that there is a constant supply of particles injected at an energy somewhat greater than the threshold energy; this is not much larger than the rest energy mc^2. The number of particles which have survived for a time t without a nuclear collision is proportional to $e^{-t/T}$ and from (6.7) their energy is approximately

$$W = mc^2 e^{t/\tau}, \tag{6.8}$$

where $B^2\tau$ is of the order of the mean time between collisions with the field. Hence $\tau\, dW = W dt$ and the number of particles in the energy range dW is proportional to $e^{-t/T}/W$ or, using (6.8) again, to $W^{-(1+\tau/T)}$. Now the observed energy spectrum of cosmic rays fits a power law well and the observed index gives 2·9 for $1+\tau/T$. Current values for B and the distance between interstellar clouds make τ an order of magnitude larger than T however, so that Fermi's original theory requires modification in order to fit the observed index. A thorough study of the change of energy of a particle in a turbulent field has yet to be made, but an interesting alternative approach will now be discussed. This is a modification of Thompson's (1955) treatment; he found a mechanism for which τ^{-1} is proportional to B as against B^2 for Fermi's mechanism.

It was pointed out in § 6.2 that the betatron mechanism is like the adiabatic compression of a gas in which the container is replaced by the field. Now, if a gas is repeatedly compressed and expanded anisotropically, some of the energy put into the compression is converted to heat owing to the irreversibility of collisions. Compression in one direction increases the component of the stress tensor belonging to that direction; then collisions redistribute the energy, so as to make the stress tensor more isotropic, and, when the expansion takes place, the stress component is smaller than during the compression and less energy is recovered than was put in. In Thompson's mechanism the collisions are replaced by the effect of irregularities in the field, so that the energy does not just go into heat, but the average gain of energy of a particle is proportional to the energy it already has, as in Fermi's mechanism.

Suppose that the field is sufficiently regular for the components $v_\parallel$ and $v_\perp$ to be meaningful. We are interested in the distribution function $f(p_\parallel, p_\perp)$, where $\mathbf{p}$ is the momentum $m\mathbf{v}$ and f is taken to be uniform in space. Since the electric field is just the induction field, it is perpendicular to $\mathbf{H}$ and does not contribute to $\mathrm{d}p_\parallel/\mathrm{d}t$, while non-uniformities in $\mathbf{H}$ do not change $|\mathbf{p}|$, but do change the angle θ between $\mathbf{p}$ and $\mathbf{H}$. The plausible assumption is now made that the effect of irregularities in H may be expressed statistically as distributing the direction of p randomly; no detailed investigation of this effect has been made however, and it will be represented in the simplest mathematical form having the required property. The Boltzmann equation is then

$$\frac{\partial f}{\partial t}+\frac{\partial}{\partial p_i}\left(f\frac{\mathrm{d}p_i}{\mathrm{d}t}\right)=\nu\left\{\frac{1}{2}\int_0^\pi f(p,\theta)\sin\theta\,\mathrm{d}\theta-f\right\}, \qquad (6.9)$$

where only the betatron effect is to be included in $\mathrm{d}p_i/\mathrm{d}t$ and the effect of irregularities in the field is represented by the right-hand side, in which ν is taken to be constant and is the effective collision frequency.

The variation of the field with time is now treated as a perturbation and the unperturbed distribution function f_0 is taken to be independent of θ. H is taken to vary as $1+\alpha\,\mathrm{e}^{\mathrm{i}\omega t}$, where α is small, and from (6.1) for the betatron effect

$$\frac{\mathrm{d}p_\perp}{\mathrm{d}t}=\tfrac{1}{2}\mathrm{i}\omega\alpha\,\mathrm{e}^{\mathrm{i}\omega t}p_\perp, \tag{6.10}$$

whence
$$\frac{\partial}{\partial p_i}\left(f_0\frac{\mathrm{d}p_i}{\mathrm{d}t}\right)=\mathrm{i}\omega\alpha\,\mathrm{e}^{\mathrm{i}\omega t}\left(f_0+\tfrac{1}{2}\sin^2\theta\frac{\mathrm{d}f_0}{\mathrm{d}p}\right).$$

The resulting perturbation f_1 of first order in α is then found from (6.9) to be

$$f_1=-\alpha\,\mathrm{e}^{\mathrm{i}\omega t}\left(f_0+p\frac{\mathrm{d}f_0}{\mathrm{d}p}\frac{\tfrac{1}{2}\sin^2\theta-\tfrac{1}{3}\mathrm{i}\nu/\omega}{1-\mathrm{i}\nu/\omega}\right). \tag{6.11}$$

Now, since the magnetic field does not change p, (6.10) gives

$$\frac{\mathrm{d}p}{\mathrm{d}t}=\tfrac{1}{2}\mathrm{i}\omega\alpha\,\mathrm{e}^{\mathrm{i}\omega t}p\sin^2\theta. \tag{6.12}$$

Then taking the real parts of (6.11) and (6.12) and averaging their product over t and θ ($\overline{\sin^2\theta}=\tfrac{2}{3}$, $\overline{\sin^4\theta}=\tfrac{8}{15}$)

$$\overline{f_1\,\mathrm{d}p/\mathrm{d}t}=-\frac{\alpha^2}{90}p^2\frac{\mathrm{d}f_0}{\mathrm{d}p}\frac{\nu}{1+(\nu/\omega)^2}, \tag{6.13}$$

and this determines the mean rate of gain of energy of particles whose total momentum is p. Its dependence on $\mathrm{d}f_0/\mathrm{d}p$ shows that this mechanism tends to spread the particles uniformly in momentum. Its dependence on ν differs from Thompson's result. He assumed that ν was large and obtained a rate of acceleration proportional to ω, but physical reasoning shows that the system should approach the reversible condition as $\nu/\omega\rightarrow\infty$, because in the limit the stress tensor must always be isotropic.

In applying (6.12) to the interstellar gas the spectrum of variations in **H** must be considered. There may well be important variations with frequency greater than ν. The value of ν is very uncertain, but, for the sake of comparison with Fermi's theory, let ν be comparable to his frequency of collisions with the field. Then the ratio of the value of τ for this mechanism to that for Fermi's is a few hundred times $(B/\alpha)^2$, and, as B is believed to be substantially less than 10^{-4}, this mechanism can be substantially faster with a modest value of α.

Fermi's theory needs further modification in order to agree with the observation that, in the range of energy up to 10^{13} eV. at least, the spectrum of the heavy nuclei in the primaries follows the same power law as that of protons. Since the cross-section for nuclear collisions is much greater for the heavy nuclei, T is much smaller, and their spectrum should follow a much higher power law. Morrison, Olbert and Rossi (1954) suggested that the cosmic rays escape from the galaxy somewhat faster than they are lost by nuclear collisions. The rate of escape is independent of the mass of the particles and would therefore lead to the same power law for protons and heavy nuclei. They suggest that the mean time for escape is about a million years, so that Fermi's value of τ must be reduced by a factor of about 6; the modified mechanism described above may be fast enough to allow this.

CHAPTER 7
SOLAR PHENOMENA

7.1. Introduction

The theoretical investigations contained in the preceding chapters have been aimed mainly at elucidating some of the problems of solar physics and solar-terrestrial relationships; these problems are discussed in this chapter and the next. The interpretation of many solar phenomena is as yet very difficult and no detailed correspondence between observation and theory must be expected. In this chapter an outline of the relevant observations will be given, with some attempts at theoretical interpretation, when this is not too speculative. Some elementary knowledge will be assumed, and the interpretation of observations will generally be assumed up to the point where hydromagnetics is involved. For detailed information the reader is referred to *The Sun* (ed. Kuiper, 1953) and to original papers.

The present time is inappropriate for speculation, because important progress on the observational side is expected in the near future from the magnetograph (see § 1.1). Whereas previously the only fields detected with certainty were in sunspots, the magnetograph detects much weaker fields. We must therefore be content often to comment: 'It will be interesting to see how this phenomenon is related to the field seen on the magnetograph.'

The sun is a star of about the average type. In white light it appears as a bright sphere with a definite surface and this surface is called the photosphere: the sharpness of the limb is due to the very rapid decrease of density at this level. The radius $R_{\odot}$ is $7 . 10^5$ km. The energy radiated by the sun is generated by nuclear processes, which are important only in a central core. Knowing the rate of radiation of energy, a fairly accurate picture of the density and temperature as functions of the distance from the centre has been worked out, on the assumption that magnetic fields are unimportant in this connexion; it is also assumed that bulk motion is unimportant except in the inner core and outer shell where convection occurs. There is no reason to believe that such a field is important and the calculation is probably sound because it agrees reasonably well with the known total mass of the sun. The impossibility of observing the interior of the sun, however, is a major difficulty in interpreting magnetic phenomena. On the other hand, the atmosphere surrounding the sun can be observed by several means.

The chemical composition at the photosphere is known quite accurately and for our purposes may be used for the whole sun. It is sufficient to know that the material is mostly hydrogen, with helium the next most abundant element, and that the abundance of all heavier elements combined is about one atom to every thousand hydrogen atoms. The hydrogen is almost completely ionized except near the level of the photosphere; the heavy elements are everywhere mostly ionized, so that the electron density in the photosphere is $\sim 10^{-3}[H]$. The temperature of the photosphere is about 5000° K., and, as remarked in § 3.4, the frequency for electrons of collisions with hydrogen atoms is then smaller than that for collisions with other charged particles, even though the concentration of the latter is a thousand times smaller. The value of σ given by the formula in § 2.7 is therefore only slightly too large, and the photospheric material may be regarded as a good conductor. Nearly all the material is neutral, however, and can move relative to the field as pointed out in § 3.4. It may be mentioned that negative hydrogen ions make a very important contribution to the opacity near the photosphere; these ions are formed by the attachment of electrons to hydrogen atoms, but the proportion of electrons so attached is very small.

7.2. Photospheric granulation and convection

When the photosphere is observed with sufficient resolving power, it is found to have a speckled appearance, consisting of bright nuclei or 'granules' on a darker background. This granulation is difficult to observe because of spurious variations of intensity due to turbulence in the earth's atmosphere and convection currents set up in the telescope by the heat of the solar radiation. Some of the best photographs of granulation were obtained seventy years ago by extraordinarily patient work. The size of the granules is often found to be ~ 4000 km., but better observations show that the granules of this size are usually clusters of smaller granules of size $\sim$ 1000 km. They are separated by about the same distance. The brightness is found to vary by a few per cent of the average brightness for the larger size and by up to 40 % for the smaller size. The granulation gradually disappears near the limb but is otherwise seen over the whole disc. The pattern of the granulation changes completely in a few minutes. The pattern does not appear to move as a whole, and therefore each granule has a life of a few minutes. Doppler velocities have been measured and are only $\sim$ 0·3 km./sec. It is believed that the granulation is due to convection in a layer slightly below the photosphere, in which case veiling by the layers above must reduce the contrast and Doppler shifts. The explanation in terms of convection will now be outlined.

Convective instability occurs when the downward temperature gradient exceeds the adiabatic value, and this occurs in the theoretical model of the solar interior both in the central core and in the outer layers. Eddington (1942) suggested that in the outer convection zone the energy was transported by the convection of hydrogen ionization energy: ionized hydrogen rising, recombining near the photosphere and the neutral hydrogen so formed falling. The ionization zone is only $\sim$ 5000 km. deep, however, and it is now believed that convective instability occurs in a zone of depth $2 . 10^5$ km. or nearly $\frac{1}{3}R_{\odot}$, and ordinary convection extending to this depth is expected to be the important factor. The Reynolds number for the convective motion itself is very large, so that this motion is probably turbulent, which may account for the short life of the granules. The theory of convection gives reasonable values for the size of the granules and the velocity of the motion, and the ratio of these is a time of about 7 minutes.

Our concern here is with the effect of a magnetic field on the convection; as mentioned in § 5.6, this has been studied by Chandrasekhar (1952). It is believed that sunspot fields inhibit the convection and this will be discussed in § 7.3. Assuming that the hydrogen ionization mechanism is not involved, the material is almost completely ionized and a typical convective motion would jumble the field. The field strength might then become strong enough to influence the motion. A convective motion is possible, which does not continually increase the field energy, but merely causes large fluctuations in the field. Consider, for instance, a motion which first transports some material from a low level to a high level, where it cools by radiation, and then returns it to the original level by the reverse of the original motion. The field energy would in general be temporarily increased, but would return to its original value, yet energy would have been transported by convection. The type of convection which would occur in the absence of a magnetic field would probably amplify any field indefinitely and it is possible that the presence of even a weak magnetic field would cause an observable difference. It will be interesting to see whether the granulation is affected by the weak magnetic fields seen on the magnetograph. Macris (1955) has suggested that the granulation does vary with the solar cycle, contrary to previous belief.

7.3. Sunspots

A sunspot consists of a nearly uniform dark 'umbra' surrounded by a less dark 'penumbra', which is darker than the normal photosphere and usually shows a fine structure of more or less radial filaments. The effective temperature of the umbra is about 25 % less than that of the normal photosphere. The penumbra is often surrounded by a ring brighter than the normal photosphere. The radius of the penumbra is about double that of the umbra and is $\sim 10^4$ km. for large, but not exceptionally large, spots. The life of spots increases with their size and for large spots is about a month. A magnetic field is always found in the umbra and for large spots is ~ 2000 gauss. During the decay of a spot the area decreases gradually, but the field strength only decreases slightly until the spot has nearly disappeared, and then drops much more abruptly. For different spots the field strength is correlated with the maximum area reached. At the centre the field is nearly vertical and its inclination varies roughly linearly with the distance from the centre, until it is nearly horizontal at the outer edge of the penumbra.

Spots appear in groups which are usually bipolar in character; a simple spot pair would consist of two spots of opposite magnetic polarity separated by a distance of the same order as the diameters of their penumbrae, and with the line joining their centres nearly parallel to the solar equator. The sun rotates from east to west and the western spot of a bipolar pair is called the 'leader' and the eastern, 'the follower'. Most spot groups are more complicated. Often a single penumbra contains two or three umbrae, but these nearly always have the same polarity. There are often several follower spots with separate penumbrae, but again the same polarity, so that the group character is still bipolar in a loose sense. Single spots or 'unipolar groups' are quite common, but are usually part of a bipolar group at some stage of their history; in this case they can be classified as leader or follower, and in fact they are usually leader spots.

Sunspots are formed in bipolar regions, whose weak fields have only recently been detected by the magnetograph; spots tend to occur early in the life of the bipolar region. A spot begins as a persistent dark space between the granules, known as a 'pore', which is not much larger than a granule. Sometimes pores increase in size and form spots, but the majority never become large enough to count as spots.

Motion in spots is indicated by Doppler shifts, which were first observed by Evershed. The observations of weak metallic spectral lines, which are formed at a low photospheric level, are interpreted as a horizontal radial outflow, which increases from zero at the centre to ~ 2 km./sec. at the outer edge of the penumbra, where it drops abruptly to zero. Observations of strong lines suggest that the direction of motion is reversed in the chromosphere.

Faculae should be mentioned here. They are bright markings seen in white as well as monochromatic light; in white light they are usually visible only near the limb where the background photosphere is darker, and they appear to be elevated slightly above the photosphere. Most faculae are associated with spot groups and have similar dimensions; some are observed at high latitudes and may be related to the polar field (see §7.5). Faculae are irregular in form, consisting of patches with holes in them. They may change in a few hours but are generally longer lived than spots. Their brightness shows that they are not cooler than the photosphere and that they must be denser than their surroundings if their composition is the same.

The observations relevant to the structure of sunspots have now been summarized and we may consider what theory has to say.

The life of sunspots is long enough for them to approximate to static equilibrium. The magnetic pressure is $\sim 10^5$ dynes/cm.^{2} and this is comparable to the gas pressure in the photosphere, so that the magnetic force must be of major importance. In the photosphere the gas pressure is mainly contributed by neutral hydrogen, and the electron pressure is roughly a thousand times less, but a few thousand kilometres deeper the electron pressure has increased to 10^5 dynes/cm.^{2}. Well above this level $H^2 \gg 8\pi p_e$ and it may be concluded that $\mathbf{j} \wedge \mathbf{H}$ is small. It is thought that the lines of force fan out so that the field strength decreases with height. The problem is complicated not only by the complex geometry of spot groups, but by the rapid change of physical conditions with depth near the photosphere. The material in the photosphere is mostly neutral hydrogen, which can move relative to the field, and mechanisms involving ionization and recombination may be important, but the diffusion of the neutral hydrogen through the ionized constituents is rather slow. If the electron density is 10^{12} cm.$^{-3}$, the frequency for hydrogen atoms of collisions with positive ions is $\sim 10^3$ sec.$^{-1}$. The thermal velocity of hydrogen atoms is ~ 10 km./sec. and, with a scale length of 1000 km., the velocity of diffusion is ~ 10 cm./sec. and the time required ~ 100 days.

The most likely explanation of the cooling of spots is the inhibition of convection by the magnetic field. It has been seen that the field is strong enough near the photosphere to be important mechanically and the absence of convection seems to be sufficient to account for the temperature drop (Sweet, 1955). The filaments seen in the penumbra are suggestive of convection cells which are forced to slope by the fanning out of the field; they are longer lived than normal granules and this would be so, if they were controlled by the field. Some of the energy which would, in the absence of a spot, be convected upward through the site of the spot, may be diverted to its edges and cause the bright ring which is sometimes seen to surround the penumbra.

Parker (1955) has made a simplified investigation neglecting the diffusion of neutral hydrogen and neglecting the 'curvature' part of the magnetic force. With these simplifications the vertical gradient of the field strength depends simply on the difference of density between the spot and the normal surroundings at the same level. Let the temperature and pressure of the spot be δT and δp less than in the surroundings. Then, if $p\delta T > T\delta p$, the density in the spot is greater than in the surroundings and H increases upwards. So long as the spot is cooler than its surroundings $\delta p/p$ decreases downwards, and it seems quite possible that H should decrease downwards below some level. It is not known to what depth the field extends but, if it causes a sufficient temperature drop, H need not increase with depth. Parker suggests that when the cooling mechanism sets in, the thermal contraction concentrates the weak field of the bipolar region; the cooling and concentration of the field enhance each other. If the cooling is sufficient, H must increase upwards rapidly, and it is possible that H could be increased by a large factor at the photosphere.

Observations of the early stages of spots are few, though they are known to start as pores. It is uncertain whether the field is already strong, when it is brought to the surface, or whether it is strengthened after reaching the photosphere, as in Parker's suggestion. The decay of spots has been better observed and it is known that the bipolar region outlives the spots. The persistence of a strong field at the centre of a spot almost throughout its life, while the area decreases gradually, shows that the decay mechanism is concentrated near the edge of the spot: the spot 'melts' round its edge. The discussion of the motion of a general field in § 3.3 shows that, when a spot is surrounded by a region of weak field and the spot decays, flux from the spot must go into the weak field (though flux from the weak field may be destroyed elsewhere at the same time). The Ohmic field would have the right effect qualitatively and would be concentrated round the edge, if the current were concentrated round the edge; the average Ohmic field would need to be greatly amplified by turbulence to make the decay rapid enough, however. Alternatively, the motion of the field might be due to motion of the material in the reverse manner to Parker's suggestion for the formation of spots.

The Evershed effect is puzzling. The length of life of spots requires the velocity of the field to be much less than 2 km./sec. and therefore any appreciable motion of the charged particles should be parallel to the field. Then the velocity should have a vertical component not much smaller than the horizontal components, except very near the outer edge, but no vertical component is observed. The inflow observed in the chromosphere is equally difficult to understand. In the absence of a magnetic field, the cooling of the atmosphere above a spot might be expected to set up such a flow, with a down flow over the spot at intermediate levels. The effect of the field could be avoided, if the field were in strands with field-free material between, in which the motion could occur. The field would have to be arranged so as to stop the flow at the edge of the spot. The dark holes in faculae might be due to a stranded field. If the gas pressure in the bright parts is *increased* above that of the surroundings by the magnetic force the 'curvature' force must be involved. This is one reason for considering whether the field is twisted.

Attention was paid in chapters 3 and 4 to the possibility of cosmic fields being twisted and it is now asked whether sunspot fields are. The inclination of the field to the vertical has been measured, but the direction of the horizontal component has not; it is usually assumed to be radially outwards, but need not be radial. The vortex structure which appears both in the penumbral filaments and on the spectroheliograph suggest that the field is appreciably twisted. Mechanical considerations show that the twisting should be either right-handed for both spots of a pair or left-handed for both and this accords with the observation that the vortices have the same sense for both. Twisting would have another mechanical effect which is essentially the same as that discussed in §4.6. On the elastic string picture, without twisting, the magnetic tension should pull the spots of a group towards each other, whereas they are observed to drift slowly apart. The discussion of §4.6 suggests that twisting might account for this.

7.4. Statistics of sunspots

The first periodicity discovered in the sun concerned the number of spots, which was found to have maxima at roughly regular intervals; the mean period is 11·1 years, but the interval between consecutive maxima ranges from 8 to 17 years. The average time from maximum to minimum is 6·6 years. A 'cycle' means the period between consecutive minima and the intensity of the maximum is found to be greater for shorter cycles.

Sunspots occur in a belt of latitude which varies with the phase of the cycle and never exceeds 40°. At minimum the belt for the new cycle is centred on about 30° and the latitude decreases to 16° at maximum, after which the belt continues to drift towards the equator; there is no lower limit to the latitude at which spots may occur. The cycles overlap somewhat so that at minimum there are belts at 30° and 5°, but few spots in either. The magnetograph has only operated near minimum, but bipolar regions of the old cycle were observed to extend to latitude 50°.

There is a simple rule for the polarity of bipolar groups, which is obeyed by 97% of all those groups which can be classed as bipolar. The polarity is the same for all groups of the same cycle in the same hemisphere, but is reversed in the opposite hemisphere and in alternate cycles. When this is taken into account the period must be the length of two consecutive cycles, and this repeats more consistently than the length of a single cycle; it varies from 18 to 26 years.

Alfvén (1950*b*) has discovered a correlation between successive cycles in opposite hemispheres, that is cycles of the same polarity. Taking the difference between spot numbers in the two hemispheres and reversing the sign of the difference for alternate cycles, he finds that the values for successive cycles are correlated.

A few theoretical comments on the solar cycle are made in §7.10.

7.5. The corona, general magnetic field and rotation

It is only at total eclipses that the corona can be observed satisfactorily in white light. Nearly all of the continuum is photospheric radiation scattered by either electrons (K component) or dust particles (F component). The F component is negligible in the inner corona but exceeds the K component at distances from the centre greater than about $2R_{\odot}$. The two components are distinguished by observations of the polarization of the light, but it is difficult to obtain accurate values of the electron density in the outer corona by this means. The intensity decreases with distance from the sun not only because of the decrease of the electron density, but also because of the decrease in the intensity of the light incident on the electrons. Because the intensity decreases rapidly with distance, a photograph can only show good contrast in a small range of this distance, the region of good contrast depending on the exposure used. For this reason many striking features of eclipses can only be seen in drawings.

The corona changes greatly with the solar cycle, having its most definite form at sunspot minimum. The form of the minimum corona is shown schematically in fig. 7.1; the regular features of fig. 7.1 are reproduced quite faithfully at each minimum. In the lowest part the intensity is almost independent of latitude, but at

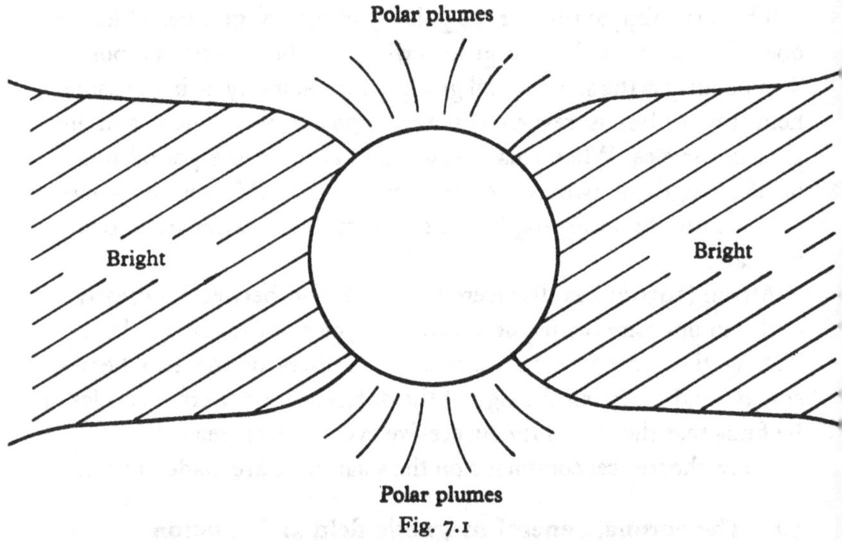

Fig. 7.1

somewhat greater heights the corona is markedly brighter at low latitudes than in the polar regions; the bright part has sharp edges which are usually at a latitude of 60–70°. Farther from the sun these edges curve away from the poles and the bright part is continued by the equatorial streamers which can be traced out to a distance of several solar radii. At maximum the corona is irregular with streamers extending in many directions, and no indication of the direction of the solar axis. The corona is brighter at maximum even when compared with the equatorial part at minimum.

The darker spaces in the polar regions of the minimum corona contain thin bright rays emerging nearly vertically from the photosphere and curving away from the solar axis. Their appearance led to the suggestion that they are related to a magnetic field, and this has been confirmed by the magnetograph. The bright rays must be tubes in which the gas pressure is not much reduced by the field and hence tubes of weak field. Van de Hulst (1950) has found that the variation of density along a ray conforms with that required by mechanical equilibrium. He also showed that at a height of $0{\cdot}2R_\odot$ the density in a ray is several times the background density. It follows that the magnetic pressure must be comparable with the gas pressure in a ray; this requires at least 1 gauss. The field seen on the magnetograph has opposite polarities in the two polar regions and is confined to latitudes greater than 55°. It is seen to be very irregular; the component in the line of sight is ~ 1 gauss. No comparison with the detail of the coronal structure has yet been made.

Theory predicts that the 'general' field just described must move with the material, and must therefore persist during sunspot maximum, though it may be transported, compressed or expanded by motion of the material. It remains to be seen what the magnetograph will show in the coming maximum. No account will be taken in the following of a general field at latitudes less than 55°. Apart from its failure to appear on the magnetograph at minimum, there is no strong evidence in favour of such a field.

The corona is remarkably hot: its temperature is roughly a million degrees, but there is a discrepancy between different methods of measurement. The intensity of radio noise under quiet conditions and the degree of ionization of iron ions both require a temperature of about $7\,.\,10^5\,°$ K., and there is no likely source of error in these methods. The variation of density with height and the width of spectral lines require a temperature greater than $1{\cdot}5\,.\,10^6\,°$ K., but the effective temperature obtained by either of these methods could be increased by irregular motion in the corona. It may be noted that the corona is seen to be 'rough', with irregular variations of density, and it is possible that there are corresponding variations of temperature, but it is difficult to avoid the conclusion that some sort of irregular motion plays an important role. Such motion is involved by both the principal mechanisms proposed for the heating of the corona. The loss of heat is about $2\,.\,10^4$ ergs/cm.^{2} sec. by conduction into the chromosphere and about half as much by radiation. In one mechanism the motion is driven by convection below the corona. The energy density of the motion is expected to be independent of height, so that the velocity is proportional to $\mu^{-\frac{1}{2}}$. When the velocity approaches the speed of sound, the motion consists of a large number of shock waves, and at this stage the rate of dissipation is large. The other mechanism involves the accretion of material from the interstellar gas. The material is mainly ionized before it arrives within $10R_{\odot}$ of the sun. The velocity of arrival of a particle falling freely in from space is a few times larger than the thermal velocity at $10^6\,^\circ$K. The fall of the accreted material is retarded by collisions with particles belonging to the normal corona, the important collisions being those between protons; the accreted material has lost most of its momentum when it has fallen to a height of $0{\cdot}2R_{\odot}$. Accretion would set up convection currents in sloping cells with velocities of the order of the velocity of sound (Hoyle, 1949).

Either of these mechanisms could be affected by magnetic fields. If there were a general field all over the sun charged particles could not cross the lines of force; a local field would prevent or divert the accretion locally. The effect on shock waves would be less drastic, but might be expected to show in the short-lived structure. It may be noted that the viscous heating is always much more important than Joule heating, because $4\pi\sigma\eta/c^2 \sim 10^5$. In any case local fields should cause local temperature differences; observations of the coronal lines (see § 7.6) do suggest that at minimum the polar corona is cooler than the equatorial part. It may therefore be possible to use the effect of local magnetic fields to distinguish between different heating mechanisms.

The rotation of the sun with its polar field at minimum presents a problem of the type discussed in § 5.2. The rotation is measured by following sunspots and other long-lived features across the disc and also from Doppler shifts observed at the eastern and western limbs. The angular velocity decreases with increasing latitude; the period of rotation varies from about 25 days at the equator to about 31 days near the poles. In the equatorial plane the centrifugal acceleration equals gravity at about $30R_{\odot}$, but the rays are not seen as far out as this. The discussion of § 5.2 should be relevant and it would be of great interest to know the form of the field in the surrounding gas, but this has not yet been achieved. The conclusion, that the rotation of tubes of force which go far into the surrounding gas would be retarded, suggests a natural explanation for the variation of the angular velocity with latitude. The extension of the corona to a distance as large as $200R_{\odot}$ (Siedentopf, Behr and Elsasser, 1953) in a disc nearly perpendicular to the solar axis also suggests a centrifugal effect. The quantities involved are very uncertain, but will now be briefly discussed.

The ordinary viscosity of the solar interior would hardly change the distribution of angular velocity in a period as long as the age of the sun, but eddy viscosity may be very much stronger. The outer convection zone is thought to be quite deep (see §7.2) and eddy viscosity here might be important. At the same time the shear flow due to the non-uniform rotation has a very high Reynolds's number and could cause some sort of turbulence even in the convectively stable region. The stability prevents any lump of material from moving far in the vertical direction, but this does not preclude all irregular motion.

The torque on the sun resulting from a twist in the magnetic field could be estimated quite accurately, but only a rough order of magnitude is of interest. If the component of **H** perpendicular to the meridian plane were comparable to the component in the meridian plane, the angular acceleration would be roughly of order $H^2R_{\odot}/M_{\odot}$. Such a strong twisting should be apparent in the polar rays and as an east-west asymmetry on the magnetograph, and is not observed; but with the twisting reduced by a factor of 10, the effect would still be important in $\sim 10^8$ years. It may, therefore, have been important in the evolution of the sun, but not on any shorter time scale.

7.6. Observations in monochromatic light

The most intense absorption lines in the solar spectrum, namely the Hα line and the H and K lines of Ca II (singly ionized calcium), are very broad and they can be seen in emission beyond the limb: The Hα line extends to the greatest height, 12,000 km., and this part of the solar atmosphere is called the chromosphere. With the spectroheliograph the sun is observed in the light of a wavelength band which can be much narrower than the total widths of these lines. As the height in the chromosphere corresponding to optical depth equals one varies with wavelength from zero in the extreme wings to several thousand kilometres at the centre, the chromosphere can in principle be observed over this range of height. The interpretation of the observations is difficult, however, because the theory of the formation of the lines is complicated and because of the irregular structure of the chromosphere.

The middle chromosphere shows marked granulation, but this is very faint in the lower chromosphere; the granules appear to be somewhat larger than the photospheric granules. At the limb the chromosphere is seen to contain bright 'spicules' of width similar to that of the granules. The spicules shoot up from the bottom with speeds ~ 20 km./sec. and are visible above the top of the background chromosphere. The relation between these phenomena is uncertain. Some form of turbulence is required to account for the slowness of the decrease of pressure with height in the chromosphere.

The spicules are frequently aligned parallel to each other over an extensive part of the limb and their direction can often be related to markings in the corona. The alignment is expected to be controlled by a magnetic field, and it is to be hoped that observations of spicules will soon be compared with magnetic regions observed by the magnetograph. If this expectation is fulfilled the direction of the spicules will be useful in helping to determine the direction of the field.

Some particular phenomena observed with the spectroheliograph are discussed in §§7.7 and 7.8, and mention should now be made of the coronagraph, invented by Lyot. In the coronagraph an image of the sun is formed and is eclipsed by a reflector in the image plane, but even under the best conditions it is very difficult to observe the corona in white light. It is much easier to observe the emission lines of the corona, using a filter. The strongest lines are regularly observed in the inner corona. They are the red line of Fe X, the green line of Fe XIV and the yellow line, whose identification is uncertain. The yellow line is seen only in disturbed regions, and its presence is believed to indicate an unusually high temperature. The red and green lines are nearly always present and their intensity varies with the solar cycle in a way similar to the white corona, but more closely related to prominences. They are usually absent from the polar regions at minimum and this is interpreted by Waldmeier as indicating a low temperature. He has also used the ratio of the intensities of the red and green lines to estimate temperatures, but it is difficult to understand how the Fe X and Fe XIV lines can be formed side by side, and it has been suggested that they are formed in different parts of a fine structure. The red and green lines in fact show different detail.

7.7. Prominences

Prominences are seen in the light of Hα or the H or K line of Ca II; at the limb they appear as bright regions above the normal chromosphere and on the disc as dark 'filaments', which are nearly always elongated as the name implies. Relative to the surrounding corona a prominence is believed to be about 100 times denser with its temperature reduced by a factor of the same order. There is great variety among prominences and there is a continuous transition between filaments and the chromospheric structure on the disc. They differ from spicules in that downward motion of the bright regions predominates, though surge prominences surge up like big spicules and then fall back. The motions are beautifully displayed in kinematograph films. In all types of prominence the motion gives the impression of being guided by a magnetic field; sometimes the paths converge to a point on the limb, usually in a spot, and sometimes bright knots are seen to move successively along the same path. Often prominences have a fine structure of bright threads. All prominences originate in the sunspot zones, though less than half originate near spot groups. Seen on the disc as filaments there is no difference between those formed near spot groups and the others. The magnetograph has shown that filaments usually lie along the borders of bipolar regions and are associated with fields of ~ 1 gauss. It is expected, therefore, that prominences are essentially magnetic phenomena. We have seen that a magnetic field causes variations of gas pressure, and it is believed that the increase in the rate of radiation of energy, due to the increased density, can account for the cooling. More detailed theoretical study is difficult because the interpretation of spectroheliograms in terms of density and temperature is itself difficult, but the long-lived 'quiescent' prominences will now be further discussed.

Quiescent prominences sometimes last for many rotations and have a fairly systematic behaviour. D'Azambuja has given typical dimensions for a fully developed quiescent prominence as $2 \cdot 10^5$ km. long, $4 \cdot 10^4$ km. high and $6 \cdot 10^3$ km. wide; such a prominence has the form of a vertical blade, but the blade is in general curved. The prominence first appears as a filament on the poleward side of a bipolar spot group. It drifts towards the poles at a rate of 1° per rotation. The direction of the filament rotates in a way which can be accounted for by the differential rotation of the sun. If the rotation of the direction of the filament is extrapolated back from the birth of the filament to the birth of the spot group, the extrapolated direction is nearly N.-S. Filaments at high latitudes are old and their directions are nearly E.-W. When a N.-S. filament is observed at the limb it appears as a series of arches like a multispan bridge. An E.-W. filament appears at the limb as a nearly vertical blade.

When there is an E.-W. filament on the limb a system of arches is seen in the white corona; the arches are one inside the other, alternately bright and dark, and their shape agrees well with the lines of force in fig. 4.4. Remembering that the intensity of the white light indicates electron density, the model of §4.3(iii) could be applied, with A chosen so that the field is strong where the dark arches occur. The width of the arches is often comparable to $2\pi h$, where h is the scale height of the corona but is occasionally larger, whereas in the model it cannot be larger. The width is also comparable to the length of the filament so that the use of a two-dimensional model is barely justified. Nevertheless, the agreement with fig. 4.4 suggests that the model contains some truth. The assumption of isothermal conditions is totally wrong for the prominence itself, but the prominence occupies a relatively small central region and it is plausible that it is caused by the increase of pressure due to the squeezing by the lines of force. It may be noted that all prominences are centred in a dark space in the white corona, which can probably be explained in the same way. Similar arch systems are also seen in the light of the coronal lines, but they are more difficult to interpret, and may involve temperature variations.

7.8. Flares

A flare is a sudden and short-lived increase of intensity best seen in H α; a few have been seen in white light. Because the occurrence of a flare cannot be predicted their beginning is rarely seen, but they are known to reach maximum intensity in a few minutes. They decay more slowly; the duration of a moderate flare is about half an hour. Their frequency of occurrence increases with decreasing magnitude down to the limit of observation; their frequency follows the spot number quite closely and is several per day at maximum. Flares occur only near spot groups: they occur more often near complex groups and more often when the spots are changing rapidly. Motion is observed in the neighbourhood of flares as a Doppler shift, and the velocity often reaches 500 km./sec. At first the velocity is upwards and decreases at a rate rather less than solar gravity until eventually it is downwards. At the limb this phenomenon appears as a surge prominence; these rise at any angle to the vertical and return along the same path. Violent motion is also often observed in existing filaments at about the time of a flare, sometimes before the flare appears.

Flares are accompanied by a number of other emissions. A few very large increases in the intensity of cosmic rays have been observed, and it is found that the bulk of the additional radiation is at the low energy end of the spectrum (usually $< 10^{10}$ eV.). These are detected most efficiently by neutron monitors, and increases are observed for most flares of large or medium magnitude, though they can only be observed in certain zones on the earth, because of the effect of the geomagnetic field. These increases are observed about half an hour after the flare.

'Outbursts' of radio noise are roughly simultaneous with flares. Ionospheric fadeouts are almost certainly due to ultraviolet emission from flares.

Sudden commencement magnetic storms (see chapter 8) occur some 30 hr. after a flare and have been attributed to a stream of ionized gas emitted at the time of a flare with a velocity of 1000 km./sec. This is known as 'corpuscular emission'.

Estimates have been made of the energy emitted in these various forms from a large flare. It is thought that the visible and ultraviolet light contain the most energy, somewhat less than 10^{30} ergs and that the corpuscular emission contains $\sim 10^{29}$ ergs. The energy of the cosmic rays is $\sim 10^{28}$ ergs and that of the radio noise very much less.

Giovanelli (1947) suggested that flares result from electric discharges occurring at neutral points of the magnetic field. He was led to this suggestion by his experience of observing flares and supported it by the correlation of flares with complex groups and with rapidly changing groups and by the positions of flares relative to the spots. We now consider whether all the phenomena associated with flares can be accounted for by the mechanism described in § 6.4. The production of soft cosmic rays is an essential feature, and the sudden onset is incompatible with Fermi's type of mechanism; it therefore seems that some sort of discharge must be involved and no other reasonable explanation has been proposed. It is suggested here that a discharge occurs at a neutral point, probably above the chromosphere, and that the visible flare is caused by the excitation of chromospheric atoms and ions by the accelerated particles, so that its position is partly controlled by the density of excitable particles. In this case the light from the accelerating region is a negligible contribution to the intensity seen by a spectroheliograph, though the strong broadening of the lines may be partly due to large bulk velocities near the accelerating region. The flare surges and corpuscular streams have velocities of ~ 1000 km./sec. and it is reasonable to assume that velocities of this order occur near the accelerating region; it is also reasonable energetically. It was seen in § 6.4 that the electric field in the accelerating region is approximately the same as the induced field a little way away. With $u \sim 1000$ km./sec., the voltage of the discharge is then $\sim aH$ volts. The values of a and H are uncertain, but may be guessed as 3000 km. and 30 gauss, and these values agree with the emission of particles with 10^{10} eV.; larger values are possible. Voltages of this order are of course likely to occur in the neighbourhood of spot groups in the absence of neutral points; it was seen in § 6.4 that a neutral point is required in order that charged particles may fall through the potential drop.

The other important quantity is the number of particles accelerated, which might be expected to be small on account of the thinness of the discharge. Since the thickness b is inversely proportional to the electron density n, however, and the number of particles accelerated is proportional to nb, this number is independent of n; it therefore does not depend strongly on the height of the accelerating region. The number of particles accelerated per second is $nabc$ or $Hac/4\pi e$. Assuming $Ha \sim 10^{10}$ gauss cm., this number is $5 \cdot 10^{28}$ particles/sec.; if these were spread over a hemisphere at the earth's distance they would be some twenty times as numerous as the normal cosmic rays.

The orders of magnitude of the time scale and the total energy released can also be roughly checked. Using the previous values the time a/u is only 3 sec., and, if a is replaced by the distance separating the spots, the sudden onset still presents no difficulty. The energy released can be an appreciable fraction of the total magnetic energy in the atmosphere above a spot group: with $H \sim 100$ gauss and $a \sim 30{,}000$ km., $a^3H^2/8\pi \sim 10^{31}$ ergs, and less than 10^{30} ergs are released. Thus the orders of magnitude are all reasonable and we next consider whether the other phenomena accompanying flares can be explained.

There are two features of the theoretical mechanism which may be responsible for secondary phenomena; the high energy particles and the bulk motion of the gas. The high energy particles should explain the increase of cosmic ray intensity observed at the earth; the observations are compatible with a spectrum that is cut off sharply above a quite low energy. The delay of the cosmic ray increase has not yet been explained, and the observation of increases associated with most flares does not agree with the theoretical suggestion of a beam that is narrow in one direction; the beam might sweep over a large range of directions, however.

The visible and ultraviolet radiation can result from the impingement of the high energy particles on a region of large enough density, in the same way that an aurora can result from a beam of high energy particles entering the earth's atmosphere from outside. The emission region would be thicker than the accelerating region because of scattering by collisions, and would in any case only appear very thin if seen end on; the observed emission region usually has a filamentary form. The brightening of parts of nearby prominences can be explained similarly by the impingement of the beam on the prominence, which has a large density.

The emission of corpuscular streams, surge prominences, Doppler shifts and the activation of prominences could result from the motion in the discharge mechanism. It was seen in § 6.4 how the lines of force are 'reconnected' at the neutral point and this happens quickly. It may be helpful to think of a mechanical analogy consisting of a system of strings; if some strings are cut and rejoined in a different arrangement the mechanical equilibrium must in general be upset and violent motion may result. In the case of flares violent motion may occur anywhere in the magnetic system of the spot group. The activation of prominences usually follows a flare with a delay which corresponds to a reasonable velocity of travel for a hydromagnetic disturbance. In the emission of corpuscular streams the buoyancy force might be important, but it is not certain whether this alone could achieve a supersonic velocity. It is important to know whether a magnetic field is carried away by the stream; it is unlikely that such a field would still be attached to the sun, because the lines of force would then be stretched, but a discharge could detach a loop of field from the main spot field. A further effect of the motion might occur if there were more than one neutral point in the group. It was pointed out in § 6.4 that a discharge does not set in until the gradient of the field in the region of the neutral point becomes large enough. Now the motion associated with one discharge may increase the field gradient at another neutral point, so that the discharge at one neutral point may trigger off some others. Flares in fact often appear to be multiple.

The radio emission could be due either to the high energy particles radiating in the magnetic field or to 'plasma' radiation due to the violent motion. This is still a controversial question. The observed spectrum favours the latter mechanism (Wild, Murray and Rowe, 1954), but the theory has not developed far and is outside the scope of this book.

7.9. *M* regions

The observation that magnetic storms (see chapter 8) often occur in series, the storms recurring at 27-day intervals, led Bartels to suggest that they were due to some unknown but long-lived solar phenomenon, which would be near the centre of the disc three or four days before the storm; he called them *M* regions. Statistics have shown that they are not correlated with prominences and are negatively correlated with the red and green coronal lines (the *M* region is on the eastern limb about ten days before the storm). *M* regions occur most frequently towards the end of a cycle and one long-lived *M* region has occurred since observations with the magnetograph began. It showed a large unipolar region of weak field: area $\sim 0{\cdot}15 \times$ disc, mean field $\sim 0{\cdot}5$ gauss. Unipolar regions occur in the same zones of latitude as bipolar regions but are relatively rare. Unlike bipolar regions they are not associated with bright plages on the spectroheliogram.

There seems little doubt that the *M* regions postulated by Bartels are unipolar regions. A seasonal effect, which may be described as an increase of recurrent storms with the magnitude of the earth's heliographic latitude, may be accounted for by the latitude distribution of unipolar regions, which is probably similar to that for spots; it is necessary to assume that the storm-producing agent travels radially out from the sun in a narrow beam, not more than a few degrees wide.

This phenomenon, as it is magnetic at both ends, is a challenge to the theory of cosmic electrodynamics, but as yet theory has had very little to say about it. It is supposed that streams of ionized gas travel from the *M* region to the earth in 3 days, which requires a mean speed of 500 km./sec., though the speed would be smaller if the stream were ejected from a region preceding the unipolar region, and would be ten times smaller, if the true time of travel were 3 days + 1 period of rotation. The stream must be ejected in a narrow cone and last for several months. No mechanism for its ejection has been proposed. Further understanding of the phenomenon may come from observations of the outer corona. It may be asked how far from the sun the magnetic field extends. It was seen in § 5.3, that at a certain distance from the solar axis ($30R_{\odot}$ or about the radius of Mercury's orbit) the centrifugal force becomes important, and that the situation cannot then approximate to a steady state. Further investigation of the behaviour when the field does extend far out may be needed.

7.10. The problem of the origin of sunspots and of the solar cycle

Because of the regular variation of the latitude and polarity of sunspots, the mechanism which produces them must be closely related to the mechanism of the solar cycle. Spots (and the bipolar regions seen on the magnetograph) almost certainly arise from the transport of a magnetic field from the solar interior to the photosphere. The problem is therefore a problem of the interior, which cannot be observed. The only field known to exist in the interior is the general field. The total flux of a large spot or large bipolar region is comparable to the flux of the general field, which tends to suggest that the general field is the source of the spot fields; the source field need not have so large a flux, however, as a tube of force may pass through the same spot any number of times. The form of the general field in the interior is not known; it may spread through most of the volume and extend nearly to the photosphere, it may be concentrated near the axis or concentrated in a tube not axially symmetric. It may be possible to deduce this form theoretically, but no attempt has yet been made. In the spread-out form the field strength would be ~ 1 gauss throughout, but in the concentrated form it could be much larger.

Several attempts have been made (the first by Walen (1949)) to account for the solar cycle by torsional oscillations similar to those described in §5.5 in connexion with variable magnetic stars. The observed polarity of spot groups and bipolar regions suggests that the lowest even mode predominates. The irregularity of the 22-year cycle has not been accounted for, but it has been suggested (Anderson, 1954) that the cycle is truly periodic with a period of 169 years. The calculated period is very uncertain, but for a plausible model is on the large side. The period of the fundamental mode is roughly the travel time for Alfvén waves from one pole to the other, and, if this is 169 years, the average reciprocal wave velocity is about 0·03 sec./cm. Allowing a field of 10 gauss, and taking the square mean root density for a solar diameter as 15 g./cm^3., the average reciprocal wave velocity is greater than 1 sec./cm. Consequently, it is necessary to assume either a stronger field, which implies that the field is strongly concentrated, or a mode of oscillation in which only the outer part of the Sun is important, or a still longer fundamental period.

The transport of the field to the surface in the spot zones is not explained. Buoyancy may be involved and the phenomenon described in § 4.5 of the formation of loops in twisted fields may be the original mechanism of the generation of spots, whether the loops break away from the general field or not. If this were involved, the spot fields would be twisted as suggested in § 7.3.

An alternative explanation assumes an additional field in the interior independent of the known general field, but this is an *ad hoc* assumption for which there is no clear need.

CHAPTER 8

MAGNETIC STORMS AND AURORAE

8.1. Magnetic storms

The geomagnetic field is continually varying and its variations are recorded at magnetic observatories, which are widely distributed over the earth. The time-scale of the records is usually about 15 mm. per hour. The records show a regular diurnal variation, but this is often dominated by less regular disturbances. When there are strong disturbances all over the world, the phenomenon is known as a magnetic storm. The average morphology of the storm field is here briefly described.

The undisturbed field approximates closely to that of a dipole near the centre of the earth, but tilted at an angle of 11° to the geographic axis. The geomagnetic poles, equator, latitude Φ and longitude are defined in terms of the axis of symmetry of this field instead of the geographic axis. In this chapter the 'geomagnetic' will be omitted, but will always be intended. The horizontal component H of the dipole field is then everywhere directed towards the north and the vertical component Z is downwards north of the equator and upwards south of it; the magnitude of Z is denoted by V. At the poles V is about 0·6 gauss and at the equator H is about 0·3 gauss.

Many storms begin with a very rapid increase in H occurring at all places with $\Phi < 60°$; the time of onset seldom differs from place to place by more than a minute. This phenomenon is known as a sudden commencement. Sudden commencement storms are always associated with solar flares and the greatest storms belong to this class. The recurring storms associated with M regions (see §7.9) do not show sudden commencements.

The average behaviour of sudden commencement storms has been analysed using the data of a large number of storms. The 'storm time' is the time elapsed after the sudden commencement, and the 'storm time variation' is the mean of the variations for the individual storms at a fixed storm time. Assuming that sudden commencements are randomly distributed in time of day, the storm time variation must be independent of time of day (or longitude) and depends only on storm time and latitude. It is found that the form of the variation is the same for all latitudes less than 60°. When the storm time variation is subtracted from the variations of the individual storms and the remainders are used to form a new average, this time at each hour of local time, another regular variation is found which is known as S_D. At high latitudes S_D is very strong and this has prevented the determination of the storm time variation for $\Phi > 60°$.

The storm time variation is much stronger in H than in either of the other two elements. It includes the sudden commencement itself which leads to a maximum in H; this might be 20γ larger than the original value ($10^5\gamma = 1$ gauss). The maximum is followed by a decrease such that H falls below its original value a few hours after the sudden commencement; the period in which H is greater that its original value is known as the 'initial phase', and the remainder of the storm, during which H is depressed, as the 'main phase'. About 10 hours after the beginning of the main phase H reaches a minimum, which is further below its original value than the maximum was above it. The subsequent recovery of H to its normal value takes many days, and all other signs of disturbance may disappear before H recovers.

The storm time variation in declination is very small. The variation in Z is several times smaller than that in H; it varies similarly with storm time, but in such a way that the magnitude V of the vertical component decreases when H increases and vice versa. Thus the variation in Z is opposite in the two hemispheres and is zero on the equator.

If sudden commencements are always worldwide, their frequency is not expected to depend on the time of day. The magnitude of sudden commencements observed at any particular observatory may, however, depend on the time of day; this would imply that the magnitude of any particular sudden commencement varies with longitude. If this were so, and if also there were some sudden commencements which were too weak to be detected, the frequency of sudden commencements observed at a particular observatory would depend on time of day, because sudden commencements in a certain range of absolute intensity would be above the critical level for detection for part of the day and below it for the other part. Newton and Milsom (1954) found that at Greenwich sudden commencements are more frequent at night than in the day. Forbush and Vestine (1955) found that at Huancayo near the equator their magnitude and also that of the initial phase was greater in the daytime, but they could not detect any variation of the frequency with time of day.

The S_D component of the disturbance may be roughly described by sinusoidal variations of period 1 day in each of the components of the field. The S_D field is strong at high latitudes where it is often $\sim 50\gamma$. Unlike the quiet day variation (see § 9.2) it is not larger by day than by night, nor particularly strong at the equator. The phases and amplitudes of the variations change rapidly with latitude near the zones of maximum auroral frequency, which are roughly the circles $\Phi = 67°$. At high latitudes H and V have their maxima in the morning and the declination varies with comparable amplitude with such a phase that the horizontal component rotates clockwise in the northern hemisphere. Near the auroral zones the variations in V and in the declination are small and the time of maximum H is near noon. At $\Phi = 60°$ H and V have their maxima in the evening and the declination again varies with its phase corresponding to a clockwise rotation in the northern hemisphere; the variation at $\Phi = 60°$ therefore differs from that at high latitudes just by a change of phase of 180°. It is sometimes found that the latitude of the reversal is lower for intense storms. At $\Phi = 50°$ the phase of H but not V goes through another reversal. If the local time of maximum H is plotted as longitude on a map, the points lie on a spiral (Meek, 1955; Hope, 1956).

For an individual storm the amplitude of S_D is weaker on successive days, but its behaviour in the initial phase is not known. The S_D variation is found to have the same form for M region storms, which do not have sudden commencements. So far as is known the form of the S_D variation is independent of the strength of the disturbance and the same is true of the storm time variation except that H attains its minimum somewhat earlier for more intense storms.

The intensity of rapid irregular variations is much greater on days of storm. On quiet days the irregular variations are much greater near the auroral zones than elsewhere.

Earth currents are observed during magnetic storms. Their morphology generally agrees with that expected from induction by the varying magnetic field.

8.2. Aurorae

Aurorae are luminous regions in the upper atmosphere, which are bright enough to be visible at night. They have a variety of forms and colours. Their spectrum contains lines and bands of the atmospheric constituents. Their forms will not be described here in detail, but the following should be noted.

Quiet arcs, appearing like a rainbow in shape, are common. A quiet arc has a sharp lower border, but no well-defined upper border; the maximum light intensity occurs a few kilometres above the lower border and decreases rapidly just above the maximum. Arcs lie nearly along parallels of geomagnetic latitude and often extend over a large range of geomagnetic longitude; as seen from a particular observing station they often reach the horizon at both ends.

Many forms show ray structure. The rays may be as thin as 1 km. They are aligned parallel to the geomagnetic field. The vertical extension of forms with ray structure is greater than that of quiet arcs and the intensity of rays may not vary greatly over a range of up to 100 km. Forms with ray structure are generally brighter but shorter-lived than quiet arcs.

A diffuse luminous surface or glow appears like a cloud over a large area of the sky without any sharp borders; its light is faint. Some glows, but not all, are the upper parts of arcs, whose intense lower part is below the horizon.

The heights of aurorae are determined by triangulation and they may extend from 90 km. to several hundred kilometres; because of their great height they are visible from large distances. It is probable that some aurorae are caused by charged particles of high energy entering the atmosphere from outside. The height of the lower-border then provides an estimate of the range of the particles in air and hence of their energy. If the lower border is at 90 km. the estimated energy is ~ 1 MeV., if the particles are protons, and ~ 4 KeV., if they are electrons. The presence of incoming protons in quiet arcs was inferred by Meinel (1951) from his observation of the Hα line with a Doppler shift. When his line of sight was perpendicular to the geomagnetic field there was no Doppler shift, but when it was parallel there was a shift corresponding to an incoming velocity of 3000 km./sec. This provides a lower limit for the velocity of the protons, when they enter the atmosphere; it is about one-fifth of the velocity corresponding to 1 MeV.

The frequency of occurrence of aurorae is found to depend strongly on latitude. It has a maximum at $\Phi = 67°$. The belts near this latitude are called the auroral zones, and aurorae are seen there nearly every clear night. Near the poles the frequency is smaller and at $\Phi = 50°$ it is only a few per year, while at low latitudes aurorae are hardly ever seen. There is a well-established relation between aurorae and magnetic storms. The lowest latitude at which aurorae are seen decreases as the strength of the storm increases. Thus the occurrence of aurorae *at moderate latitudes* is positively correlated with storms. During a storm an aurora may move over a large range of latitude during a night, so that it is seen over a very wide area, but during some storms strong aurorae are not seen in the auroral zones.

Since the frequency of storms is much less than the frequency of aurorae in the auroral zones, aurorae do occur under magnetically quiet conditions, and it is important to discover whether there is any fundamental difference between the aurorae associated with magnetic storms and those occurring nearly every night in the auroral zones. It has been found that the total luminosity of aurorae is positively correlated with the intensity of magnetic disturbance and that forms with ray structure are more likely to be accompanied by magnetic disturbance than those without. From observations in the auroral zones alone it is also found that the brighter aurorae are on the average lower, but at lower latitudes the average height of the lower border is the same as that for all forms in the auroral zones, though the vertical extension is greater.

The frequency of aurorae varies with local time and has a maximum at about 23.00 hr. Forms with ray structure occur more frequently before 23.00 hr. than after. Aurorae can also be observed during the day by radio techniques; with frequencies higher than 20 Mc./s. rapidly fading echoes are obtained, which are believed to result from scattering by raylets of ionization, many times thinner than the rays. The frequency observed with this technique has maxima before and after midnight; it is very small between sunrise and noon and increases gradually through the afternoon.

The intensity of the visible radiation from an aurora has been estimated by Hulburt (1928) as 10^{-2} ergs cm.$^{-2}$ sec.$^{-1}$. The energy flux of the incident particles must be greater because not all the energy is radiated as visible light and also because the cross-section of the aurora perpendicular to the field is substantially less than the effective radiating area. The ionization associated with aurorae led Sugiura, Tazima and Nagata (1952) to an estimate of the density of the primary beam of 1 proton per c.c., which for 1 MeV. protons corresponds to 10^3 ergs cm.$^{-2}$ sec.$^{-1}$.

Particles with energy intermediate between the auroral primaries and cosmic rays have been detected at heights of 50 km. and higher by observations with rockoons by Meredith, Gottlieb and van Allen (1955). They suggested that these particles belong to a high energy tail on the spectrum of auroral primaries, because they were found at latitudes 64° and 74° but not at 55·6° or 88·5°. The intensity was ~ 10 particles cm.$^{-2}$ sec.$^{-1}$. The intensity varied considerably over distances smaller than the radius of the orbits for 10 MeV. protons and they concluded that the particles are probably electrons.

8.3. The Chapman–Ferraro model

If the correlation between sudden commencement storms and solar flares is to be explained by a direct connexion between the two phenomena, it is necessary to suppose that the agent responsible for the storm travels from the sun to the earth in a time of ~ 30 hr.; its average speed is then ~ 1000 km./sec. Chapman and Ferraro suggested that a stream of ionized gas arrives in the vicinity of the earth with a speed of this order of magnitude and have made several theoretical investigations of the interaction between such a stream and the earth's magnetic field. The earlier suggestion of a stream of charged particles of one kind only was shown by Lindemann to be untenable, because the electric field would be large; considerations like those of § 1.3 show that the densities of positive and negative particles must be very nearly equal. Chapman and Ferraro assumed that there was no magnetic field in the stream when it left the sun, but Hoyle (1949) and Alfvén (1950*a*) have proposed modifications in which there is such a field. It is not yet certain whether this modification enables their models to account for more observations, but it makes the theoretical problem much more complicated, and these theories have not been developed in detail. Here the simplest and most important results of Chapman and Ferraro's work will be presented.

This problem differs from those of chapter 5 in that collisions are not sufficiently frequent to maintain an approximately thermal distribution. A fundamental investigation is therefore required, but it is useful to consider first the simplified problem in which the stream is a perfectly conducting fluid.

(i) *Perfectly conducting stream.* If a perfectly conducting stream leaves the sun with no magnetic field in it, the result of § 3.2 shows that no field can exist in the stream at any subsequent time. Suppose that initially the stream has a plane front surface, which advances towards a fixed magnetic dipole. Since the field is confined to the space outside the stream, the lines of force near the front surface are pushed towards the dipole. It is assumed that there is no current outside the stream and then the disturbance in the field for a plane surface can be represented by an image dipole as in fig. 8.1. The field just outside the stream is finite and the field changes discontinuously to zero at the surface, so that there is a finite surface current in this idealized model.

The magnetic pressure acting on the front surface of the stream opposes its advance and, if the dipole is roughly parallel to the surface, the pressure is greatest on the part of the surface near P. The magnetic pressure therefore distorts this part of the surface, so that it becomes concave towards the dipole. As the surface advances towards the dipole the magnetic pressure increases and the advance of the surface is eventually stopped, when the magnetic pressure equals the pressure exerted by the stream on the surface. Far from the dipole the field can have little effect on the motion of the stream. Consequently the stream flows round on all sides of the dipole, and the surface is folded round the dipole to form a cavity in the stream; the field is confined to the inside of the cavity by currents in the surface. Mechanically the cavity behaves as an elastic obstacle in the stream.

Since the shape assumed by the surface is unknown, it is convenient to use the field of fig. 8.1 to obtain the orders of magnitude of the quantities involved. If distances are measured in units of the earth's radius and R_0 is the distance from the centre of the earth to P, the disturbance field H_D at the centre of the earth is $0{\cdot}3\,(2R_0)^{-3}$ gauss, and the total field just outside the stream at P is $16H_D$. Taking the pressure of the stream on the surface at P to be

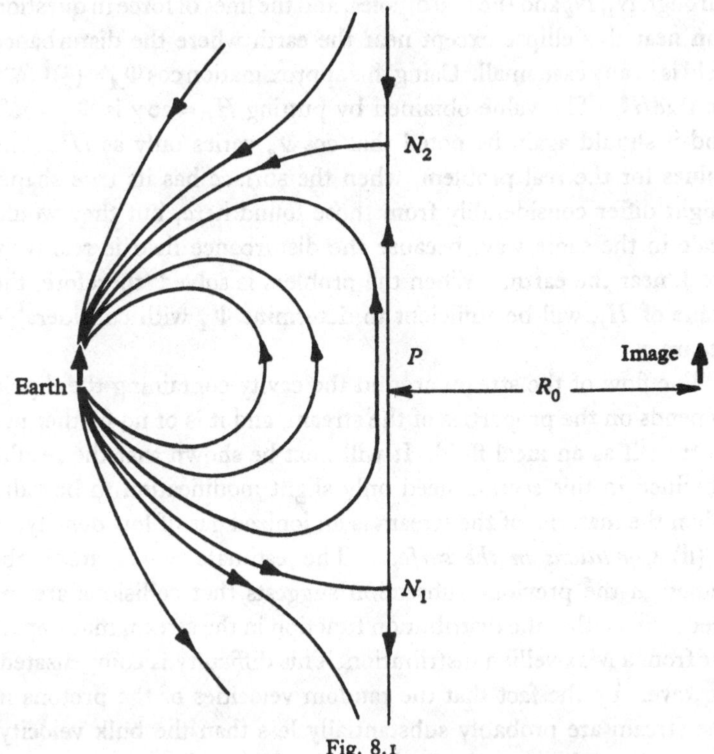

Fig. 8.1

$2nm_p U^2$, where n is the particle density in the stream and $U \sim 1000$ km./sec., and equating this to the magnetic pressure at P gives $n \sim 3.10^8 H_D^2$ cm.$^{-3}$. The field H_D increases H at low latitudes as observed in the initial phase of a storm. A typical value of H_D for the initial phase is 20γ giving $R_0 = 6$ and $n \sim 12$ cm.$^{-3}$; it should be noted that R_0 varies only as $H_D^{-\frac{1}{3}}$, while $n \propto H_D^2$.

An important feature of the field is the existence of two neutral points N_1 and N_2 on the surface; $N_1P = 2^{-\frac{1}{2}}R_0$. Every line of force on the surface starts at N_1 and ends at N_2. The importance of these neutral points is that charged particles from the region occupied by the stream can travel down the lines of force from N_1 and N_2 to the earth, and it is therefore of interest to know the latitude Φ_A at which these lines of force reach the ground. These lines of force deviate little from the lines of force of the earth's dipole alone; the locus on which the fields of the two dipoles are parallel is an ellipse through N_1, N_2 and the two dipoles, and the lines of force in question run near this ellipse except near the earth where the disturbance field is in any case small. Using this approximation $\cos\Phi_A = (\frac{2}{3})^{\frac{3}{4}} R_0^{-\frac{1}{2}}$ or $1{\cdot}28 H_D^{\frac{1}{6}}$. The value obtained by putting $H_D = 20\gamma$ is $\Phi_A = 72°$, and it should again be noted that $\cos\Phi_A$ varies only as $H_D^{\frac{1}{6}}$. The values for the real problem, when the surface has its true shape, might differ considerably from those found here, but they would scale in the same way, because the disturbance field is relatively weak near the earth. When the problem is solved, therefore, the value of H_D will be sufficient to determine Φ_A with considerable accuracy.

The flow of the stream around the cavity containing the dipole depends on the properties of the stream, and it is of no further use to treat it as an ideal fluid. It will next be shown that the results obtained in this section need only slight modification to be valid when the material of the stream is an ionized gas of low density.

(ii) *Conditions in the surface.* The estimate $n \sim 12$ cm.$^{-3}$ obtained in the previous subsection suggests that collisions are infrequent, so that the distribution function in the stream may depart far from a Maxwellian distribution. This difficulty is compensated, however, by the fact that the random velocities of the protons in the stream are probably substantially less than the bulk velocity, because 1000 km./sec. is the thermal velocity of protons at about 10^8° K. It is therefore reasonable to consider next the approximation in which both the random velocities and collisions are neglected. When the field is known, the problem then reduces to the calculation of the orbits of the particles, the initial velocity being the same for all. It may be remembered from § 1.5 that it is difficult to solve cosmic electrodynamical problems solely by calculating orbits, because the field usually depends on the orbits. In the present case, however, it will be assumed that the field is almost the same as that found in (i). This can be justified by the discussion of § 2.7;

the only effect of the departure from thermal conditions is in the stress terms of (2.16) and these were seen to be small. It will therefore be assumed that the field is very weak in the main body of the stream and that the stream is excluded from a cavity containing the dipole. The field must now penetrate slightly into the stream, because it must deflect the orbits of the particles away from the cavity. The electric current in the stream is easily calculated when there is no random motion and hence the required modification of the field can be found. The penetration of the field into the stream will now be investigated, neglecting the curvature of the surface; this approximation is justified because the field is found to be important only in a layer which is thin compared to the radius of curvature. Consider first a steady state.

Let the field have everywhere the z direction and depend only on x, the x-axis being normal to the surface and pointing into the stream. To be consistent with this field all quantities referring to the stream must depend on x only. The orbit of a particle entering the surface layer, where the field is important, is bent round until it emerges again into the main body of the stream, where the field is weak. Because of their greater momentum the protons tend to penetrate further into the field than the electrons, but only a slight charge separation is needed to set up an electric field sufficient to balance this tendency. This is a Hall electric field and it is normal to the surface. An upper limit to the potential difference belonging to this field is set by the initial kinetic energy of the protons. In a steady state $\mathbf{E} = -\nabla\phi$ and, since ϕ is a function of x only, $\mathbf{E}$ is normal to the surface. Then the orbits of the particles are symmetrical between their incoming and outgoing parts, and, since there is no force in the z-direction, the velocity of one kind of particle can be written $(-u(x), v(x), W)$ for ingoing particles and $(u(x), v(x), W)$ for outgoing particles, where W is constant. The density $n(x)$ is also the same for incoming and outgoing particles.

Since u, v, $W \ll c$, relativistic terms may be neglected. The magnetic field may be represented by a vector potential having only a y-component $A(x)$. The limiting values of n, u and v as $x \to \infty$ will be denoted by N, U and V. Then the generalized integral of momentum gives

$$v_p + eA/m_p c = v_e - eA/m_e c = V, \tag{8.1}$$

where the suffices refer to protons and electrons, the conservation of energy gives

$$u_p^2+v_p^2+2e\phi/m_p=u_e^2+v_e^2-2e\phi/m_e=U^2+V^2, \qquad (8.2)$$

and the equation of continuity gives

$$n_p u_p=n_e u_e=NU. \qquad (8.3)$$

From (8.2) we note

$$\tfrac{1}{2}m_p(U^2+V^2)\geqslant e\phi\geqslant -\tfrac{1}{2}m_e(U^2+V^2). \qquad (8.4)$$

Eliminating ϕ from (8.2) and using (8.1) we have

$$m_p u_p^2+m_e u_e^2+(m_p^{-1}+m_e^{-1})(eA/c)^2=(m_p+m_e)U^2, \qquad (8.5)$$

whence

$$A^2\leqslant m_p m_e c^2U^2/e^2=A_0^2. \qquad (8.6)$$

The field must now be made consistent with the current and charge density in the stream. Since the outgoing stream differs from the ingoing stream only in the sign of u, the inhomogeneous Maxwell equations give

$$\frac{d^2A}{dx^2}=-\frac{8\pi e}{c}(n_p v_p-n_e v_e) \qquad (8.7)$$

and

$$\frac{d^2\phi}{dx^2}=-8\pi e(n_p-n_e). \qquad (8.8)$$

Now substituting from (8.1) for v_p and v_e in (8.7)

$$\frac{d^2A}{dx^2}=\frac{8\pi e^2A}{c^2}\left(\frac{n_p}{m_p}+\frac{n_e}{m_e}\right)-\frac{8\pi eV}{c}(n_p-n_e). \qquad (8.9)$$

It has already been stated that n_p and n_e must be nearly equal in the main body of the stream, but this needs verification for the surface layer. Suppose that $n_p\approx n_e$; then from (8.3) $u_p\approx u_e$, and (8.1) and (8.2) give

$$\phi\approx VA/c+eA^2/2m_ec^2, \qquad (8.10)$$

where terms in m_e/m_p are neglected. Then an expression for $d^2\phi/dx^2$ can be obtained using (8.9) and, provided $(dA/dx)^2$ is not greater than $A\,d^2A/dx^2$, the order of magnitude of $d^2\phi/dx^2$ is $16\pi ne^2\phi/m_ec^2$. Hence using (8.8) and (8.4)

$$\frac{|n_p-n_e|}{n}<\frac{m_p(U^2+V^2)}{m_ec^2}, \qquad (8.11)$$

and with $(U^2+V^2)^{\frac{1}{2}}\sim 1000$ km./sec., n_p and n_e will not differ by more than 2 %. This argument applies to any part of the stream, provided $(\mathrm{d}A/\mathrm{d}x)^2 < A\,\mathrm{d}^2A/\mathrm{d}x^2$, which will be verified.

The inequality (8.11) shows that the ratio of the second term on the right-hand side of (8.9) to the first term is at most $(m_p/m_e)^{\frac{1}{2}}\,UV/c^2$ and, when the second term is neglected, the equations can be integrated. From (8.3) and (8.5)

$$n = N(1-(A/A_0)^2)^{-\frac{1}{2}}. \tag{8.12}$$

Writing $x'^2 = 8\pi Ne^2x^2/m_ec^2$ integration of (8.9) then gives

$$\frac{1}{2}\left(\frac{\mathrm{d}A}{\mathrm{d}x'}\right)^2 = A_0^2\{1-(1-(A/A_0)^2)^{\frac{1}{2}}\}, \tag{8.13}$$

the physical boundary condition being $\mathrm{d}A/\mathrm{d}x \to 0$ as $x \to \infty$, and the additional condition $A \to 0$ as $x \to \infty$ being chosen. Putting $A = A_0 \sin\theta$, integration of (8.13) gives

$$x' = -\log\tan\tfrac{1}{4}\theta - 2\cos\tfrac{1}{2}\theta + \text{arbitrary constant}, \tag{8.14}$$

the sign being chosen so that $x \to +\infty$ as $\theta \to 0$. The surface of the stream is given by $u = 0$, and then from (8.5) $A = A_0$ and hence $\theta = \frac{1}{2}\pi$. The field strength is obtained from (8.13) as

$$H = H_0 2^{\frac{1}{2}} \sin\tfrac{1}{2}\theta, \tag{8.15}$$

where

$$H_0^2 = 16\pi N m_p U^2. \tag{8.16}$$

It may now be verified that $(\mathrm{d}A/\mathrm{d}x)^2 < A\,\mathrm{d}^2A/\mathrm{d}x^2$. From (8.9) $\mathrm{d}^2A/\mathrm{d}x'^2 = A_0\tan\theta$ and hence

$$\frac{(\mathrm{d}A/\mathrm{d}x')^2}{A\,\mathrm{d}^2A/\mathrm{d}x'^2} = \frac{4\sin^2(\frac{1}{2}\theta)}{\sin\theta\tan\theta} = 1 - \tan^2\tfrac{1}{2}\theta.$$

The approximation $n_p \approx n_e$ is therefore justified and the formulae obtained here describe the field in the stream fairly accurately.

H is plotted against x' in fig. 8.2. When $\theta \ll 1$, H varies like $\mathrm{e}^{-x'}$ to a good approximation, and the field must become unimportant in the stream at a distance from the surface of a few times $(m_ec^2/8\pi Ne^2)^{\frac{1}{2}}$, which is the plasma wavelength, and is about 1 km. for $N = 20\,\text{cm.}^{-3}$.

It is now clear that the surface layer of the stream, in which the field is important, is thin compared to the dimensions of the cavity.

It is then sufficient in the case of the steady state to regard the particles as being specularly reflected at the surface, because of the symmetry between the outgoing and incoming parts of the orbits. When the surface is moving, the effect of the field on the orbits can be described as specular reflexion in a frame in which the surface is instantaneously at rest; the velocity of the surface does not change appreciably in the time in which an individual particle is in the surface layer, since this is only a few milliseconds. The pressure of the stream on the surface is, from (8.16) $2Nm_p U^2$, where U is the

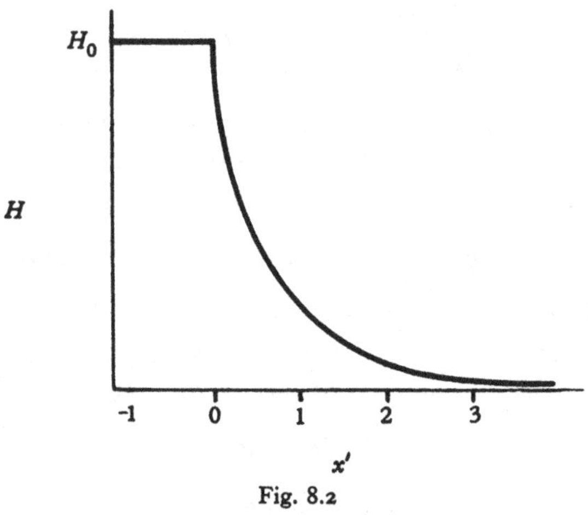

Fig. 8.2

normal component of the velocity of the stream relative to the surface. The surface must move with such a velocity that this pressure balances the magnetic pressure.

It should be noted that U, V and W vary over the surface of the cavity, because the axes were related to the field and the normal to the surface. A_0 varies proportionally to U; this expresses the fact that the particles penetrate farther into the field at normal incidence than oblique incidence. Because A_0 varies over the surface, the surface where $u=0$ is not strictly parallel to the lines of force, but the distance by which it deviates is of the same order as the thickness of the surface layer, and is therefore small. The real problem in which the stream approaches a dipole can now be formulated in a simple way. The field is completely determined by the condition that its strength at the surface should be proportional to $\cos\chi$, where χ is the angle between the normal to the surface and the velocity of the stream. Further progress seems to require the solution of this problem, but it is difficult, owing to the surface being unknown, and it is not even clear how to set about computing a solution.

(iii) *Random velocities and collisions.* The inclusion of random velocities much smaller than U would not be expected to make much difference to the model of the previous subsection, though it would make it more realistic by removing the singularity in n. They have a more important effect in closing the cavity on the 'shadow' side of the earth. In the absence of any random motion the pressure of the stream on the surface would vanish wherever its normal was perpendicular to the direction of the stream, and the cavity would have a cylindrical extension behind the earth. The random motion provides a pressure on the surface whatever its direction, and consequently the cavity is closed; its extension behind the earth depends on the magnitude of the random velocities.

In discussing collisions it is important to distinguish between collisions between particles which are both in the in-going or both in the out-going stream and collisions between particles belonging to the two different streams. Collisions between particles of the same stream can only affect the random velocities and are therefore unimportant. Collisions between electrons are also unimportant, because the protons have nearly all the energy. The electrons are dragged into the surface layer by the Hall field, which in general gives them a much larger kinetic energy than they have initially. The results obtained in (ii) are only slightly affected by the initial value of u_e and the mean value of v_e is not affected by collisions between electrons, because it is the same for the in-going and out-going streams. Collisions between protons and electrons can be described by a conductivity, and hence by a motion of the field into the stream. This is equivalent to the trapping of stream particles in the cavity; a particle which makes a collision with another of opposite charge in the surface layer is likely to be trapped in the field. The relative velocity between the material and the field is $c^2/4\pi\sigma b$, where b is $H/|\operatorname{curl}\mathbf{H}|$ and, with $b \sim 1$ km., is only $3(10^4 T^{-1})^{\frac{3}{2}}$ m./s. Collisions between protons of the different streams would be important, if their frequency were appreciable. The mean free path for these collisions is roughly

$$(n \log (10^{10}))^{-1} (m_p U^2/e^2)^2 \text{ cm.} \quad \text{and} \quad U \sim 10^8 \text{ cm./sec.};$$

this is $\sim 10^{20} n^{-1}$ cm., and so is much greater than the radius of the earth's orbit. All kinds of collision are therefore unimportant in this problem.

(iv) *The horns.* Attention was drawn in (i) to the neutral points N_1 and N_2 of fig. 8.1 and corresponding neutral points must exist on the surface, whatever its shape. In the steady-state model with random velocities neglected, so that the field strength $\propto \cos \chi$, χ must be 90° at these points. When random velocities are included, the resulting pressure must indent the surface near these points; the stream can intrude into regions which are shaped like tubes of force from N_1 and N_2 to the earth. Chapman called these regions 'horns'. The conditions in a horn have not been investigated in detail, but it is unlikely that the field is excluded from the horns except in the part near N_1 or N_2. It is certain, however, that the possibility of a charged particle going to any depth in the horn is not restricted by any absolute limit of energy. The perturbation method for orbits shows that a low energy particle starting from deep in a horn will come out with its velocity nearly parallel to the field, and since the orbit is reversible this is also the condition for deep penetration. The possibility of particles of auroral energy going from outside the cavity down the horns into the atmosphere may well be significant for the theory of aurorae, but unfortunately the relation between Φ_A and H_D for the true model is not known.

8.4. Interpretation of magnetic storms and aurorae

The theoretical discussion of magnetic storms and aurorae will be based on Chapman and Ferraro's model; though the model may require substantial modification, the essential ingredient of a mechanical disturbance originating near the sun is widely accepted. The initial phase of a storm can be explained by the disturbance field of fig. 8.1, except that the observed variation in V is small at all latitudes. This may be accounted for by currents induced inside the earth (Chapman and Bartels, 1940), which oppose any change of the field inside the earth. Since the time of growth of the sudden commencement is much smaller than the time of travel of the stream, the C–F explanation requires that the particles leave the sun with a very small spread in their speed.

The decrease in H during the main phase has not been satisfactorily explained. Stormer proposed that it is due to a ring of current in the equatorial plane. While this provides a concise description of the field, no reason has been given why the current should be concentrated in a ring, and the C–F model suggests that it is more likely to be spread over a surface. The long life of a current ring, calculated from the conductivity, is often stressed, but the mechanical stability of the current is of paramount importance.

The S_D field has not been explained. The disturbance field of the C–F model has a component depending on local time, but this cannot fit the observations; for instance, the lines of force in the polar regions should be bent away from the sun, whereas they are observed to be bent towards the morning side of the earth. The rapid spatial variation in the S_D field near the auroral zones requires the current there to be relatively near the earth, almost certainly in the ionosphere. One means of generating a current in the ionosphere is a wind in the neutral air molecules (see §9.2), but it is extremely difficult to see how a wind could be generated in anything similar to the C–F model except by a current, so that this mechanism carries us no further forward. It therefore seems most likely that the current is due to an electric field in the ionosphere; the problem is complicated because of the importance of the Hall effect.

The conditions envisaged in the C–F model should exist all the time, because the orbital motion of the earth through the surrounding ionized gas is equivalent to the motion of a stream of gas past the earth. The density of the gas in the neighbourhood of the earth varies considerably but is several hundred electrons per c.c. (Siedentopf, Behr and Elsasser, 1953). The earth's velocity is 20 km./sec. and the velocity of the gas relative to the sun must be neglected, as it is unknown. Taking $N = 500$ cm.$^{-3}$ the model of fig. 8.1 gives $H_D = 2.6\gamma$, $R_0 = 11$ and $\Phi_A = 77°$.

It has been noted that aurorae occur nearly every night in the auroral zones, and that quiet arcs, which are common, show Doppler shifted Hα lines, indicating that they contain incoming protons. In a field of 0·4 gauss the radius of the orbit of a 1 MeV. proton is

3·5 km., and with the C–F model such particles cannot reach the earth except by coming down the horns. The auroral zones could therefore be accounted for, if under magnetically quiet conditions the true value of Φ_A were 67° instead of 77° given by the model of fig. 8.1. The discrepancy cannot be removed by changing the values of N or U as $\cos\Phi_A$ varies only as $(NU^2)^{\frac{1}{12}}$. It is possible, however, that the discrepancy is due to the use of the model of fig. 8.1 instead of the true field, which has not been worked out. If so, the enhancement of the irregular magnetic variations in the auroral zones may also be explained, if they are due to small disturbances in the surrounding gas, which tend to travel down the horns.

Since quiet arcs occur without magnetic storms, it seems that the primary auroral particles must be always present in the interplanetary region. The most important effect of a magnetic storm is to move the auroral zone to a lower latitude. The calculation of § 7.3 (ii) gave $\Phi_A = 72°$, so that there is a serious discrepancy in this calculation, but this does not affect the truth of the preceding statement. The observed correlation between the intensity of aurorae and magnetic storms suggests that the intensity of the primaries is increased by a storm. If the energy flux of the primaries is 10^3 ergs cm.$^{-2}$ sec.$^{-1}$ as obtained by Sugiura, Tazima and Nagata (1952), it is doubtful whether they can originate in a flare, because multiplying by the square of the radius of the earth's orbit gives $2 . 10^{28}$ ergs/sec., which is greater than the rate of release of energy in a flare. Another suggestion due to Hoyle (1949) is that there is a magnetic field in the stream and that the auroral primaries are accelerated at the neutral points in the field by the mechanism of § 6.4. With $U \sim 1000$ km./sec., we obtain $E \sim H$ volts/cm. Then, if the length of the discharge is 10^4 km., a field of 100γ is sufficient to obtain particles of energy 1 MeV. With this mechanism aurorae would occur where the lines of force from the neutral points reach the earth and their latitude decreases with increasing strength of the disturbance field in a way similar to that of the C–F model.

The previous considerations suggest that an aurora should be limited to a small area of the earth's surface and this is often true. On the other hand, arc forms extend over a large range of magnetic longitude and the magnetic characteristics of the auroral zones do not refer only to a small part of the zone at a time. The possibility should therefore be considered, that the lines of force from the whole zone 'connect with outer space' all the time or at least that the active part of the zone is elongated along the zone. A little speculation on this question will be made at the end of this chapter.

8.5. The outer atmosphere

The behaviour of magnetic storms may be influenced by currents induced inside the C–F cavity. The effect of the ionosphere has been studied by Ashour and Price (1948), using an idealized model in which the ionosphere is treated as a thin conducting sheet, whose conductivity is greater on the sunlit side of the earth. They found that the ionosphere tends to screen the disturbance field from the lower atmosphere and that, since the screening is less efficient at night, the frequency of sudden commencements should be greater at night, as observed by Newton but not by Forbush and Vestine. Ashour and Price neglected the Hall effect, but this may not be justified, as it is known to be important in other ionospheric problems (see §9.2).

The region between the level of maximum electron density in the ionosphere and the surface of the C–F cavity will be called the 'outer atmosphere'. This region has been studied (Dungey, 1955), though not specifically in connexion with magnetic storms, and its properties will be outlined. The densities of both neutral particles and ions are sufficiently low for collisions to be of little importance. Consequently the outer atmosphere approximates closely to diffusion equilibrium; that is the force of gravity on each component is approximately balanced by its partial pressure, though in the case of charged particles, there is an important force due to the electric field which is needed to keep the total charge density zero. Above a certain level, whose height is not accurately known, the velocity distribution of neutral particles becomes unsymmetrical, because an appreciable proportion of them exceed the velocity of escape. Charged particles are trapped by the field, however, and set up an isothermal distribution, with a temperature which varies, but is $\sim 1000^\circ$ K. If there were only one kind of positive ion of mass m_p the density of ions or electrons would vary as $\exp[-m_p\Phi/2kT]$, where Φ is the gravitational potential. An estimate of the electron density has been obtained using Storey's (1953) interpretation of the dispersion of whistling atmospherics and is a few hundred electrons per c.c. This value is compatible with the maximum value in the ionosphere, only if the positive ions are predominantly hydrogen ions. That the electron density in the outer atmosphere is comparable to that in interplanetary space is not predicted by any theoretical considerations and appears to be a coincidence.

The pressure of the ionization is $\sim 10^{-10}$ dynes cm.$^{-2}$, and this is less than the magnetic pressure when $H > 5\gamma$, which is true throughout the C–F cavity. The mechanical behaviour is therefore controlled by the field and the component of $\mathbf{j}$ perpendicular to $\mathbf{H}$ must be small. Unless a current flows down the whole length of a line of force in the outer atmosphere it follows that $\mathbf{j}$ is small, and this is generally true; in particular the neglect of currents in the outer atmosphere is justified for the C–F problem. The gas pressure on the inside of the surface is an order of magnitude smaller than the pressure of the stream on the outside, which is therefore mainly balanced by the magnetic pressure, as they assumed.

The relative motion between the field and the charged particles is generally negligible. The motion of the field is strongly influenced by the electric field in the lower ionosphere, where the neutral atoms control the motion. For instance, the law of isorotation can be justified, so that the outer atmosphere rotates with the earth. The velocity of rotation at $R = 11$ is 5 km./sec., which is comparable with the thermal velocity of protons at 1000° K., but is only a quarter of the earth's orbital velocity.

The behaviour of disturbances in the outer atmosphere may be discussed in terms of Alfvén waves. In the weakest part of the field $H \sim 40\gamma$ and the gyrofrequency for protons is 4 sec.$^{-1}$. For waves of much lower frequencies, therefore, the Hall effect is negligible and the waves of opposite circular polarization have nearly equal velocities (see § 5.41). The most important cause of attenuation is viscosity; the collision frequency for protons colliding with protons is ~ 1 per minute and the attenuation in one wavelength in the region of weakest field is not large for periods greater than 1 min. (Dungey, 1954). The velocity of Alfvén waves, when $H = 3 . 10^{-4}$ gauss and $N = 300$ cm.$^{-3}$ is 40 km./sec. The corresponding time of travel for a distance of 10 earth's radii is 25 min. The outer atmosphere may therefore be expected to have resonant modes of oscillation of the type described by standing Alfvén waves, and the period of the lowest mode should be $\sim\frac{1}{2}$ hr. The modes of the torsional type discussed in § 5.5 have been computed for a model with a dipole field, and it is found that a disturbance of 1γ at the ground requires a displacement near the C–F surface of 3000 km., so that the magnetograph deflexions must be small for free oscillations. The micropulsations observed with periods of ~ 1 min. are sufficiently regular to suggest that a resonant system is involved. They are often observed over a large part of the earth and it is tempting to identify them with high order modes of the outer atmosphere. A very difficult feature to explain, however, is that they are observed simultaneously in different regions with different periods.

8.6. Waves in the C–F surface

Since there is a sudden change in the bulk velocity of the material at the C–F surface, it may be expected that surface waves will develop, like water waves generated by a wind. The stability problem has been worked out (Dungey, 1955) using a plane model and instability is found for waves whose component of velocity parallel to the 'wind' in the stream is positive, but smaller than the 'wind' velocity. There is a preference for the wavefronts to be parallel to $\mathbf{H}$, as well as a preference for them to be perpendicular to $\mathbf{u}$. Their general direction of travel is therefore eastward by day and westward by night, though motion of the interplanetary gas relative to the sun could change the time of reversal by several hours, if its velocity were as large as 20 km./sec. This kind of behaviour is observed in the motion of nearly all irregularities in the F layer and also for the motion of rays in aurorae (Meinel and Schulte, 1953), but waves in the C–F surface certainly cannot account for all of these. The waves can be propagated into the ionosphere both as Alfvén waves and as sound waves travelling parallel to the field; except in the outer part of the outer atmosphere, the field is so strong that the tubes of force can be regarded as rigid in discussing these sound waves. In the case of Alfvén waves, assuming that the displacement of the C–F surface is less than the wavelength, it is found that the magnetic disturbance at the ground is only a few gamma, and that the corresponding disturbance in the electron density in the ionosphere is negligible. The sound wave can have appreciable amplitude in the F region; because the tubes of force are shaped like horns, they concentrate the sound wave as its travels downward. It is heavily attenuated for periods less than about 5 min., however, and since it is generated far out and travels down tubes of force, it cannot be important at low latitudes.

In discussing the C–F problem in § 8.3 it was assumed that there was a steady solution to the problem and it has now been found that the steady state is unstable with respect to the formation of waves. This instability adds to the difficulty of the steady-state problem, because it means that it cannot be solved by a variation method in which the total energy is minimized, unless some artificial modification is made to remove the instability. Because of the instability the question also arises whether the steady-state solution would be a satisfactory model for the average field. If the waves are very violent, tubes of force may be detached from the cavity like spray from water waves and would tend to be carried to the far side of the earth. Particles from space could then penetrate the earth's field at the join between such tubes and the main field, and then auroral activity could occur in regions other than those provided by the Chapman horns: these regions might be spread round a substantial part of the auroral zones.

CHAPTER 9

IONOSPHERIC ELECTRODYNAMICS

9.1. Conduction in the ionosphere

The study of the cosmic problems of the previous chapters may be helpful in the investigation of electrodynamic problems of the ionosphere, some of which are briefly discussed in this chapter. The conclusions of § 1.4 concerning the causal relationships are not generally appropriate, however, and inductance is not usually important. The ionospheric problems do not all have the same causal structure; each must be considered anew. The problems are more precise than solar problems. The undisturbed magnetic field is known and the disturbance in the field is never large. The electron density is quite well known from radio observations; it is never much larger than 10^6 cm.$^{-3}$ and the ionosphere may be defined so that it is not smaller than 10^3 cm.$^{-3}$. Equation (1.1) is not a valid approximation, but it will be seen that the order of magnitude of E is still uH/c, and it is seen that the densities of positive and negative particles are nearly equal; since $H \sim 0{\cdot}4$ gauss, $\rho/ne \sim 2.10^{-3}u/na$ and $u \sim 10^4$ cm./sec., $a \sim 10^4$ cm., $n \sim 10^3$ cm.$^{-3}$ would be extreme values.

Neutral molecules are much more numerous than charged particles and it is therefore necessary to discuss conduction in a weakly ionized gas. The Hall effect is of major importance in some ionospheric problems. The fundamental equations are still the equations of transfer of momentum for each constituent, which are expressed by (2.10), but the collision term must now be expressed in terms of collision frequencies in the way explained in § 2.5. The effects of collisions with all the different constituents must be summed, so that

$$\left[\int_{-\infty}^{\infty} \mathbf{v}\,\delta' f\,d\mathbf{v}\right]_i = n_i \sum_j \frac{m_j \nu_{ij}}{m_i + m_j}(\mathbf{u}_j - \mathbf{u}_i).$$

Several constituents are important in the ionosphere, but the approximation will be made here that only one neutral constituent and one type of ion are important in addition to the electrons. Then $n_p \approx n_e$. The most important constituents are N_2 and N_2^+ in the E

region and O and O^+ in the F region. The following numerical values will be used (Cowling, 1945; Nicolet, 1953)

$$\frac{\nu_{p,n}}{n_n} = \frac{\nu_{n,p}}{n_p} = 7 \,.\, 10^{-10}\,\text{cm.}^3\,\text{sec.}^{-1} \quad \text{for} \quad N_2^+ \text{ and } N_2$$

$$= 10^{-9}\,\text{cm.}^3\,\text{sec.}^{-1} \quad \text{for} \quad O^+ \text{ and } O,$$

$$\frac{\nu_{e,n}}{n_n} = \frac{\nu_{n,e}}{n_e} = 1{\cdot}1 \,.\, 10^{-8}(T/300)^{\frac{1}{2}}\,\text{cm.}^3\,\text{sec.}^{-1}.$$

Although $\nu_{e,p}$ exceeds $\nu_{e,n}$ in the F region, collisions between the charged particles will generally be neglected, and for convenience ν_e and ν_p will be written for $m_n \nu_{e,n}(m_n + m_e)^{-1}$ ($\approx \nu_{e,n}$) and $m_p \nu_{p,n}(m_n + m_p)^{-1}$ ($\approx \frac{1}{2}\nu_{p,n}$). Thus $\nu_e \approx 20\nu_p$ and $m_p \nu_p \approx 1500 m_e \nu_e$.

If the equation representing Ohm's law were derived in the same way as (2.16), the effects of the neutral particles would appear in the way described in § 3.4 and in particular there would be a contribution to the electric field, which would involve $\mathbf{u}_n$. It is more convenient, however, to eliminate $\mathbf{u}_n$ between the equations of transfer of momentum for electrons and positive ions. If ν_p is neglected in comparison with ν_e and $m_e \nu_e$ in comparison with $m_p \nu_p$ (though such approximations must not be made until all subtractions have been carried out), the result may be written

$$\mathbf{E} = \mathbf{j}/\sigma_0 + \mathbf{j} \wedge \mathbf{H}/nec - \mathbf{U} \wedge \mathbf{H}/c - (n_e e)^{-1} \nabla p_e - m_e \nu_e \mathbf{g}/e\nu_p + (m_e/e\nu_p)(\nu_e \, d\mathbf{u}_p/dt - \nu_p \, d\mathbf{u}_e/dt), \quad (9.1)$$

where $$\sigma_0 = n_e e^2/m_e \nu_e$$

and $$\mathbf{U} = (m_p \nu_p + m_e \nu_e)^{-1}(m_p \nu_p \mathbf{u}_p + m_e \nu_e \mathbf{u}_e).$$

The last three terms in (9.1) are generally negligible. This can be tested by comparison with the induced field; the gravitational term has the most definite value and with $H = 0{\cdot}4$ gauss,

$$m_e \nu_e gc/e\nu_p UH \sim 3 \,.\, 10^{-3} U^{-1}$$

with U in cm./sec. This shows that the gravitational term is small and the last three terms in (9.1) will now be neglected. The equation then takes the familiar form except for the difference between $\mathbf{U}$ and $\mathbf{u}$ defined in § 2.6, but in (9.1) this difference represents only a change in the Hall field of order $m_e \nu_e / m_p \nu_p$ and terms of this order have already been omitted. (Both **u** and **U** are heavily weighted in favour of the positive ions.) The effect of the neutral particles appears only in σ_0, and it must be stressed that both **U** and **u** are independent of $\mathbf{u}_n$. It is convenient to regard the whole gas as the superposition of a totally ionized gas on a gas of neutral molecules, and **u** is the velocity corresponding to the momentum density of the ionized gas.

When he first worked out the conductivity, Cowling (1945) took the induced field to be $-\mathbf{u}_n \wedge \mathbf{H}/c$. He also assumed that the inertial terms were negligible in the equations of transfer of momentum. Omitting also the gravitational term and pressure gradient the equation of transfer of momentum for electrons is

$$m_e \nu_e(\mathbf{u}_e - \mathbf{u}_n) = -e(\mathbf{E} + \mathbf{u}_e \wedge \mathbf{H}/c). \tag{9.2}$$

The equation obtained by eliminating **E** is then just

$$n_e(m_p \nu_p + m_e \nu_e)(\mathbf{U} - \mathbf{u}_n) = \mathbf{j} \wedge \mathbf{H}/c, \tag{9.3}$$

which states that the electromagnetic force density is exactly balanced by that due to collisions with neutral particles. (Usually $\rho E \ll |\mathbf{j} \wedge \mathbf{H}|/c$ in the ionosphere.) Substitution of (9.3) into (9.1) gives

$$\mathbf{E} = \mathbf{j}/\sigma_0 + \mathbf{j} \wedge \mathbf{H}/n_e ec + \mathbf{H} \wedge \mathbf{j} \wedge \mathbf{H}/n_e m_p \nu_p c^2 - \mathbf{u}_n \wedge \mathbf{H}/c, \tag{9.4}$$

which may be described as the addition of $H^2/n_e m_p \nu_p c^2$ to the direct resistivity perpendicular to the field. The physical reason for this is that the electromagnetic force causes the charged particles to move relative to the neutral particles, and the motion is resisted by collisions which dissipate energy. It must be stressed, however, that (9.3) is valid only if the pressure gradient, the force of gravity and the acceleration of the charged particles are all small compared to the electromagnetic force. This is true for some problems and it is sometimes required to solve (9.4) for j. This is most simply achieved by solving the equations for $\mathbf{u}_e$ and $\mathbf{u}_p$ separately. Thus (9.2) gives

$$m_e \nu_e(\mathbf{u}_e - \mathbf{u}_n)_\parallel = -eE_\parallel \tag{9.5}$$

$$m_e(\nu_e^2 + \Omega_e^2)(\mathbf{u}_e - \mathbf{u}_n)_\perp = -e\nu_e(\mathbf{E} + \mathbf{u}_n \wedge \mathbf{H}/c)_\perp + e\Omega_e(\mathbf{E} + \mathbf{u}_n \wedge \mathbf{H}/c) \wedge \mathbf{h}, \tag{9.6}$$

where **h** is the unit vector in the direction of **H**. Combining these equations with the corresponding equations for positive ions gives

$$j_\parallel = \sigma_0 E_\parallel, \tag{9.7}$$

$$\mathbf{j}_\perp = \sigma_1(\mathbf{E} + \mathbf{u}_n \wedge \mathbf{H}/c)_\perp + \sigma_2(\mathbf{E} + \mathbf{u}_n \wedge \mathbf{H}/c) \wedge \mathbf{h}, \tag{9.8}$$

where $$\sigma_1 = n_e e^2 \left(\frac{m_e \nu_e}{(m_e \nu_e)^2 + (eH/c)^2} + \frac{m_p \nu_p}{(m_p \nu_p)^2 + (eH/c)^2} \right) \tag{9.9}$$

and $$\sigma_2 = -n_e e^3 H/c[((m_e \nu_e)^2 + (eH/c)^2)^{-1} - ((m_p \nu_p)^2 + (eH/c)^2)^{-1}]. \tag{9.10}$$

It is seen from (9.9) that, as n_n varies, σ_1/n_e has two maxima of approximately equal value near the values of n_n corresponding to $\nu_e = \Omega_e$ and $\nu_p = \Omega_p$. In the ionosphere these occur near the heights 80 and 140 km. respectively. According to the convention used in (9.8) σ_2 is negative, because the second term in the brackets is always smaller than the first. The identity

$$\frac{\sigma_2}{\sigma_1} = \frac{\Omega_p \nu_e - \Omega_e \nu_p}{\nu_e \nu_p + \Omega_e \Omega_p} \tag{9.11}$$

will be useful later. It also shows that σ_2/σ_1 has a maximum value of about 20 at about 110 km.

The approximations to be made in treating those problems where (9.7) and (9.8) are valid may now be briefly discussed; in contrast to cosmic problems it is assumed that $\mathbf{E}$ is not nearly equal to $-\mathbf{u}_n \wedge \mathbf{H}/c$. Since the space charge density and the disturbance in the magnetic field are small, it is plausible that $\operatorname{div} \mathbf{j}$ and $\operatorname{curl} \mathbf{E}$ should be negligibly small. The former is equivalent to neglecting the displacement current and requires $E/4\pi j$ to be small compared with the time scale t. The latter depends on the disturbance in the magnetic field, which is $\sim 4\pi aj/c$ and requires $4\pi a^2 j/Ec^2 \ll t$ or $Ec^2/4\pi ju^2 \gg t$. Now low in the ionosphere σ_0, σ_1 and σ_2 are all of the order of $n_e ec/H$ or $\sim 40 n_e$ sec.$^{-1}$. This shows that the first approximation is amply satisfied, and, with $(u/c)^2 \sim 10^{-14}$, the latter is satisfied for $t \ll 10^5$ sec. With increasing height σ_0 increases, but σ_1 and σ_2 never greatly exceed $n_e ec/H$ and, when the prime cause of the current is induction, the magnitude of $\mathbf{j}$ depends on σ_1 and σ_2.

9.2. Induction by tidal motion

Even before the discovery of the ionosphere Balfour Stewart suggested that the regular diurnal variations of the geomagnetic field were due to currents induced by tidal motion of conducting air in the geomagnetic field. The magnitude of the variations is

$\sim 10\gamma$. Later, the theory of atmospheric tides showed that the tidal velocity in the ionosphere should exceed that at the ground by a factor of ~ 100, but this factor is sensitive to the way in which the atmospheric temperature varies with height and is rather uncertain (Wilkes, 1949). The use of ionospheric data to calculate the magnetic variations provides an independent estimate of this amplification factor. The theory was worked out independently by Hirono, Martyn and Fejer. The outline presented in this section follows Fejer's (1953) treatment closely.

The tilt of the geomagnetic axis relative to the geographic axis will be neglected and the tide will be symmetrical with respect to the equatorial plane. Spherical polar co-ordinates will be used, θ being the colatitude and ϕ the longitude. Equations (9.7) and (9.8) are valid and may be written in the tensor form

$$j_i = \sigma_{ij} E'_j, \tag{9.12}$$

where
$$\mathbf{E}' = \mathbf{E} + \mathbf{u}_n \wedge \mathbf{H}/c.$$

The components of σ_{ij} referred to the spherical polar co-ordinate system are required. If χ is the angle of dip, they are

$$\left.\begin{aligned} \sigma_{rr} = \sigma_0 \sin^2\chi + \sigma_1 \cos^2\chi, \quad \sigma_{\theta\theta} = \sigma_0 \cos^2\chi + \sigma_1 \sin^2\chi, \\ \sigma_{\phi\phi} = \sigma_1, \quad \sigma_{\theta r} = \sigma_{r\theta} = (\sigma_1 - \sigma_0)\sin\chi\cos\chi, \\ \sigma_{\phi r} = -\sigma_{r\phi} = \sigma_2 \cos\chi, \quad \sigma_{\theta\phi} = -\sigma_{\phi\theta} = -\sigma_2 \sin\chi. \end{aligned}\right\} \tag{9.13}$$

It is next assumed that the current is confined to a layer, which is thin compared to the earth's radius R_E. The current density below this layer must be small, because the lower atmosphere is a good insulator. Above the ionosphere the considerations of §8.5 show that the current must flow along the lines of force, but such a current is ruled out by symmetry about the equatorial plane. Now from the vanishing of $\operatorname{div}\mathbf{j}$ we have

$$j_r = -(r^2 \sin\theta)^{-1} \int_{r_0}^{r} \left(\frac{\partial}{\partial\theta}(\sin\theta\, j_\theta) + \frac{\partial j_\phi}{\partial\phi} \right) r\, dr, \tag{9.14}$$

where r_0 refers to the bottom of the current-bearing layer. Now $r \sim R_E$ but $r - r_0 \ll R_E$; then, since the integrand $\sim rj$, $j_r \ll j_\theta, j_\phi$, so that $\mathbf{j}$ is nearly horizontal. This condition determines E_r; it is expressed by

$$\sigma_{rj} E'_j = 0, \tag{9.15}$$

and is used to eliminate E_r' from (9.12), giving

$$\left.\begin{aligned} j_\theta &= \sigma'_{\theta\theta} E'_\theta + \sigma'_{\theta\phi} E'_\phi, \\ j_\phi &= \sigma'_{\phi\theta} E'_\theta + \sigma'_{\phi\phi} E'_\phi, \end{aligned}\right\} \tag{9.16}$$

where

$$\left.\begin{aligned} \sigma'_{\theta\theta} &= K^{-1}\sigma_1\sigma_0, \\ \sigma'_{\theta\phi} &= -\sigma'_{\phi\theta} = -K^{-1}\sigma_2\sigma_0\sin\chi, \\ \sigma'_{\phi\phi} &= K^{-1}\sigma_1(\sigma_0\sin^2\chi + \sigma_3\cos^2\chi), \end{aligned}\right\} \tag{9.17}$$

and

$$K = \sigma_1\cos^2\chi + \sigma_0\sin^2\chi,$$

$$\sigma_3 = \sigma_1 + \sigma_2^2/\sigma_1.$$

A further equation is obtained from (9.14) by putting $r = r_1$, which refers to the top of the current-bearing layer, for j_r must then vanish again. Since $r_1 - r_0 \ll r_0$, the factor r can be omitted from the integrand and then

$$\frac{\partial}{\partial\theta}(\sin\theta J_\theta) + \frac{\partial J_\phi}{\partial\phi} = 0, \tag{9.18}$$

where

$$J = \int_{r_0}^{r_1} j\,\mathrm{d}r.$$

The magnitudes of the disturbance in **H** and of the tidal velocity show that **E** is approximately irrotational. Now E_r determined by (9.15) may vary rapidly in the vertical direction, but not in any horizontal direction and then, if **E** is to be irrotational, E_θ and E_ϕ must not vary rapidly in the vertical direction. Although in reality $\mathbf{u}_n$ varies with height, this variation has not been included in any study of the currents, and the model is idealized by taking $\mathbf{u}_n$ and hence also E'_θ and E'_ϕ to be independent of height. Then

$$\left.\begin{aligned} J_\theta &= \Sigma_{\theta\theta} E'_\theta + \Sigma_{\theta\phi} E'_\phi, \\ J_\phi &= \Sigma_{\phi\theta} E'_\theta + \Sigma_{\phi\phi} E'_\phi, \end{aligned}\right\} \tag{9.19}$$

where

$$\Sigma = \int_{r_0}^{r_1} \sigma'\,\mathrm{d}r.$$

Another radical idealization is now made; it is assumed that n_e and hence Σ_{ij} are independent of the longitude (it is also assumed, quite reasonably, that n_n is independent of θ and ϕ). A semidiurnal tide is considered and we put

$$\mathbf{u}_n \wedge \mathbf{H}/c = \mathbf{F}(\theta)\,\mathrm{e}^{2i(\phi - t)},$$

and

$$\mathbf{E} = -r_0\nabla(\psi(\theta)\,\mathrm{e}^{2i(\phi - t)}),$$

where **F** and ψ are complex and t is the time in units of $(2\pi)^{-1}$ days. Then substituting (9.19) into (9.18) gives

$$\frac{\mathrm{d}}{\mathrm{d}\theta}\left(\sin\theta\Sigma_{\theta\theta}\frac{\mathrm{d}\psi}{\mathrm{d}\theta}\right)+2\mathrm{i}\frac{\mathrm{d}\Sigma_{\theta\phi}}{\mathrm{d}\theta}\psi-\frac{4\Sigma_{\phi\phi}}{\sin\theta}\psi$$
$$=\frac{\mathrm{d}}{\mathrm{d}\theta}(\Sigma_{\theta\theta}F_\theta+\Sigma_{\theta\phi}F_\phi)+2\mathrm{i}(\Sigma_{\phi\theta}F_\theta+\Sigma_{\phi\phi}F_\phi). \quad (9.20)$$

To complete the theory it is necessary to assume a form for **F**. Fejer assumes a form based on the observed barometric tide, but this will not be discussed here and the procedure will just be described. When $\mathbf{F}(\theta)$ is known apart from a numerical factor, (9.20) must be solved for ψ subject to the boundary conditions: $\psi=0$ when $\theta=0$ (because $E_\phi=-2\mathrm{i}\psi/\sin\theta$) and $\mathrm{d}\psi/\mathrm{d}\theta=0$ when $\theta=\frac{1}{2}\pi$ (because of the symmetry about the equatorial plane). In practice the solution is obtained by computing ψ_1, which satisfies (9.20) with $\psi_1=0$ and $\mathrm{d}\psi_1/\mathrm{d}\theta=0$ at $\theta=\frac{1}{2}\pi$, and ψ_2, which satisfies (9.20) with the right-hand side replaced by zero and $\psi_2=1$ and $\mathrm{d}\psi_2/\mathrm{d}\theta=0$ when $\theta=\frac{1}{2}\pi$. Both ψ_1 and ψ_2 diverge when $\theta\to 0$ and the required solution is $\psi_1+\alpha\psi_2$, where α is a constant chosen to make $\psi=0$ at the smallest value of θ to which the computations are carried.

Once ψ is known, **E** and **J** can be calculated, all these quantities being proportional to the unknown numerical factor in $\mathbf{u}_n$. This factor is then determined by comparison with the observed magnetic variations. In this way Fejer concludes that for the solar tide the tidal velocity in the ionosphere is about sixty times that at the ground with the same phase; he finds that the phase of the lunar tide in the ionosphere is opposite to that at the ground.

It should be remarked that the magnetic variations are much weaker by night than by day, presumably because the electron density in the E layer is much smaller by night. To include this effect in the theory it is necessary to abandon the assumption that Σ_{ij} is independent of ϕ and the computation is then much more complicated. An important unsettled question concerns the situation near sunrise: there is a rapid variation of n_e with ϕ and it is not known whether this causes a rapid variation of **E**; such an effect could also be important at an eclipse of the sun.

One striking feature of the observations, which is explained by the theory, is the sharp increase in the strength of the magnetic variations as the geomagnetic equator is approached. This may be explained as follows. In the integrated conductivities the most important contributions come from heights where $\sigma_0 \gg \sigma_1$, and in the E region σ_3 is substantially greater than σ_1. The denominator K is $\sim\sigma_0$ at most latitudes, but, as the equator is approached χ becomes small, and, when $\tan^2\chi \sim \sigma_1/\sigma_0$, K decreases rapidly, and consequently $\Sigma_{\theta\theta}$ and $\Sigma_{\phi\phi}$ increase rapidly. This occurs at latitudes between 5 and 10°, and here J_ϕ increases rapidly; J_θ vanishes at the equator because of the symmetry and does not show this effect.

Though the current-bearing layer does not extend into the F region the tidal electric field must do so, the potential being constant on a line of force. This field must cause a motion of the charged particles in the F region with a velocity which approximates closely to $c\mathbf{E}\wedge\mathbf{H}/H^2$. Since the scale of variation of $\mathbf{E}$ is $\sim R_E$ this velocity does not vary appreciably with height even in the F region. Evidence of such a motion can be seen in the behaviour of the F layer, but owing to the uncertainty of the rates of diffusion and of loss of electrons, the velocity has not yet been determined. The F layer has much larger day-to-day variations than the E layer; this suggests that there are important day-to-day changes in $\mathbf{u}_n$ in the E region.

It seems that the magnetic variations can be explained satisfactorily by currents flowing in the E region; if this explanation is accepted, the magnetic variations can be used to supplement other methods of studying winds in this region.

9.3. The motion of irregularities in the electron density

Irregularities in the electron density in the ionosphere give rise to fading of the radio echoes, and are observed to be nearly always present. Simultaneous observations with spaced receivers show that the variation in time of the irregularities includes a statistically significant mean velocity (Briggs and Spencer, 1954) and it is therefore interesting to consider how an irregularity would move, if there were a uniform motion of the neutral particles and a uniform electric field. A solution of a special problem of this type has been obtained by Clemmow, Johnson and Weekes (1955) and will be described at the end of this section. First a more general discussion will be given, but throughout this section two-dimensional symmetry will be assumed such that no quantity varies in the direction parallel to $\mathbf{H}$. In this case the disturbance in the magnetic field is parallel to the undisturbed field, but this disturbance field is neglected and the disturbance in the electric field is assumed to be irrotational. For simplicity we work in the frame in which $\mathbf{u}_n = 0$; any velocity in any other frame is then obtained by adding $\mathbf{u}_n$ to the corresponding velocity in this frame.

The behaviour of the irregularities is determined by the vanishing of div $\mathbf{j}$; putting $\mathbf{E} = -\nabla\phi$ and using (9.8) this gives

$$\operatorname{div}(\sigma_1 \nabla\phi) = \mathbf{h}.(\nabla\phi \wedge \nabla\sigma_2). \tag{9.21}$$

Then, remembering that σ_1 and σ_2 are proportional to n_e and neglecting the spatial variation of n_n, (9.21) and (9.6) give

$$m_e(\nu_e^2 + \Omega_e^2)\,\partial n_e/\partial t = e(\Omega_e + \nu_e\,\sigma_2/\sigma_1)\,\mathbf{h}.(\nabla n_e \wedge \nabla\phi). \tag{9.22}$$

We first seek a solution such that the irregularity moves without changing its form (though the particles which constitute the irregularity are continually changing); this is expressed by

$$\partial n_e/\partial t = -\mathbf{V}.\nabla n_e, \tag{9.23}$$

where $\mathbf{V}$ is a uniform velocity perpendicular to $\mathbf{H}$. For the consistency of (9.22) and (9.23) ϕ must have the form defined by

$$\nabla\phi = -\mathbf{F} + n_e^{-1}\nabla\psi(n_e), \tag{9.24}$$

where $\mathbf{F}$ is a uniform field and ψ is an arbitrary function of n_e, the n_e^{-1} in (9.24) being inserted for later convenience. Comparison of (9.22) and (9.23) requires, using the identity (9.11),

$$\mathbf{F} = -\frac{\mathbf{V} \wedge \mathbf{H}}{c}\left(1 + \frac{\nu_e \nu_p}{\Omega_e \Omega_p}\right). \tag{9.25}$$

Substituting (9.24) into (9.21) gives

$$\nabla^2\psi = (\mathbf{F} + (\sigma_2/\sigma_1)\,\mathbf{F} \wedge \mathbf{h}).\nabla n_e. \tag{9.26}$$

This means that the potential ψ is that due to a polarization which has the fixed direction of the vector in brackets and whose intensity is proportional to n_e. If the irregularities are weak, they may be treated as a perturbation δn_e, and ψ is of first order in δn_e. From (9.22) the contribution involving ψ to $\partial n_e/\partial t$ is then of second order. Consequently in the linear approximation all irregularities move without change of form with the velocity **V** determined from (9.25) with **F** equal to the unperturbed electric field. **V** is independent of the form of the irregularities; it is a weighted mean of $\mathbf{u}_e$ and $\mathbf{u}_p$.

Consider now the general case. Integration of (9.26) over a volume V bounded by a surface S of constant n_e shows that $\int_V \nabla^2\psi \,\mathrm{d}V$ vanishes. Now multiply (9.26) by ψ and integrate over the same volume. The right-hand side vanishes because ψ is a function of n_e, and the left-hand side transforms as follows

$$\int_V \psi\nabla^2\psi\,\mathrm{d}V = \psi_S \oint_S \nabla\psi\,.\,\mathrm{d}\mathbf{S} - \int_V (\nabla\psi)^2\,\mathrm{d}V$$

$$= \int_V (\psi_S\nabla^2\psi - (\nabla\psi)^2)\,\mathrm{d}V = -\int_V (\nabla\psi)^2\,\mathrm{d}V,$$

and hence

$$\int_V (\nabla\psi)^2\,\mathrm{d}V = 0. \tag{9.27}$$

This is possible only if $\nabla\psi = 0$, and then from (9.26) n_e can vary only in one specific direction in space; for such a stratified model the surfaces of constant n_e are planes. Since planes are not closed surfaces the argument just given is in any case invalid; but if n_e varies only with x, E_y must be the same everywhere, and (9.22) shows that $\partial n_e/\partial t$ is proportional to E_y; therefore any irregularity of this type moves without change of form.

The conclusion drawn from (9.27) is that no irregularity can move without change of form except in the linear approximation and except for stratified irregularities, but the model of Clemmow, Johnson and Weekes (1955) is another exception. This is an irregularity consisting of a circular cylinder in which n_e is uniform, but different from the value outside the cylinder. Then ∇n_e vanishes except at the boundary of the cylinder, and ψ need not be a function of n_e, so that the argument leading to (9.27) is invalid.

Consider any axially symmetric model and let r and θ be cylindrical co-ordinates referred to the axis of symmetry and right-handed with respect to **H**. Then (9.21) becomes

$$\frac{\partial}{\partial r}\left(r\sigma_1\frac{\partial\phi}{\partial r}\right) + \frac{\sigma_1}{r}\frac{\partial^2\phi}{\partial\theta^2} = -\frac{\partial\sigma_2}{\partial r}\frac{\partial\phi}{\partial\theta}. \tag{9.28}$$

The solution can be written in the form $\phi = \sum_m g_m(r)\,e^{im\theta}$ and separate equations are then obtained for each g_m. It is required that the electric field should be finite both on the axis and at infinity, and this is possible only for $m = 1$; then the limiting field at infinity is uniform. Putting $\phi = g(r)\,\mathrm{e}^{\mathrm{i}\theta}$ (9.28) becomes

$$\frac{\mathrm{d}}{\mathrm{d}r}\left(r\frac{\mathrm{d}g}{\mathrm{d}r}\right) = \frac{g}{r} - \frac{\mathrm{d}\log n_e}{\mathrm{d}r}\left(r\frac{\mathrm{d}g}{\mathrm{d}r} + \mathrm{i}\frac{\sigma_2}{\sigma_1}g\right). \tag{9.29}$$

With this form of ϕ (9.22) takes the form of (9.23), but $\mathbf{V}$ varies with r. Since $E_\theta = -\mathrm{i}\phi/r$, (9.22) shows that $\partial n_e/\partial t$ is proportional to $r^{-1}\phi\, \mathrm{d}n_e/\mathrm{d}r$ and on a cylinder of constant r has the value given by (9.23) and (9.25) with $\mathbf{F}$ equal to the uniform field corresponding to the potential ϕ on this cylinder. The occurrence of an imaginary term on the right-hand side of (9.29) means that the direction of $\mathbf{V}$ varies with r and hence the irregularity will in general develop some kind of spiral structure.

In Clemmow, Johnson and Weekes's model $\partial n_e/\partial r$ vanishes except at the surface of the cylinder and so the irregularity moves with the value of $\mathbf{V}$ corresponding to the surface. Inside the cylinder $\mathbf{E}$ is uniform and has the corresponding value of $\mathbf{F}$ given by (9.25). The problem of relating this to the limiting field $\mathbf{E}^0$ at infinity is analogous to the similar problem for a dielectric cylinder. If the ratio of n_e inside the cylinder to that outside is λ, the continuity of j_r across the surface requires E_r just outside the cylinder to be $(-\mathrm{i}\lambda + (\lambda - 1)\,\sigma_2/\sigma_1)F_\theta$. Putting $g(r) = Ar + Br^{-1}$ outside the cylinder,

$$A = -\tfrac{1}{2}(E_r - \mathrm{i}E_\theta)\,e^{-\mathrm{i}\theta} = \tfrac{1}{2}(\mathrm{i}(\lambda+1) - (\lambda-1)\,\sigma_2/\sigma_1)F_\theta\,\mathrm{e}^{-\mathrm{i}\theta},$$

and hence, taking real parts

$$\mathbf{E}^0 = \tfrac{1}{2}(\lambda+1)\mathbf{F} + \tfrac{1}{2}(\lambda-1)(\sigma_2/\sigma_1)\mathbf{F}\wedge\mathbf{h}. \tag{9.30}$$

This, together with (9.25), relates $\mathbf{V}$ to $\mathbf{E}^0$ (in Clemmow, Johnson and Weekes's paper, equation (3), V_x and V_y should be interchanged). When either $\nu_e \gg \Omega_e$ or $\nu_p \ll \Omega_p$, $\sigma_2 \ll \sigma_1$; in the former case the motion is strongly resisted by collisions with the neutral particles and the velocity $\mathbf{V} \approx 2(\lambda+1)^{-1}(\Omega_e\Omega_p/\nu_e\nu_p)\,c\mathbf{E}^0\wedge\mathbf{h}/H$ is small; in the latter case the motion is controlled by the magnetic field and $\mathbf{V} \approx 2(\lambda+1)^{-1}c\mathbf{E}^0\wedge\mathbf{h}/H$. When $\nu_e\nu_p \sim \Omega_e\Omega_p$, $\sigma_2 \sim 20\sigma_1$ and, unless λ is nearly equal to 1,

$$\mathbf{V} \approx -2(\lambda-1)^{-1}(\Omega_p/\nu_p)\,c\mathbf{E}_0/H.$$

9.4. Convective diffusion

In an ionospheric layer the charged particle gas is far from mechanical equilibrium, particularly in the observable part of the layer, where the electron density increases upwards. In the F region n_n is small enough for the diffusion of the charged particles through the neutral gas to be important but, since $\nu_p \ll \Omega_p$, diffusion across the magnetic field is very slow. Then, since both gravity and the partial pressure gradient are nearly vertical, it is at first sight expected that near the geomagnetic equator diffusion will be unimportant; the charged particle gas should be supported by the electromagnetic force. It was realized by Johnson and Hulburt (1950), however, that this situation is unstable and that a convective motion of the charged particles must occur; if there are irregularities in n_e, the parts where n_e is larger than normal will fall, and, since n_e increases upwards, the irregularities will be amplified. Because σ_0 is large, the electric field must correspond to a motion for the magnetic field, which is the same as the convective motion of the charged particles. Then, since the field cannot be appreciably disturbed and the spatial variation of **H** is slow, the velocity **u** must be constant on a line of force, and $\operatorname{div} \mathbf{u} \approx 0$ (there is no cause for motion parallel to **H**); then, if we follow a lump of ionization through the motion, n_e remains constant, apart from the production or loss of ionization, which are neglected here. It also follows, that, although the small initial irregularities may have any form, the irregularities generated by the motion do not vary along a line of force.

The discussion so far can apply only to a limited stretch of the lines of force, because, if they are followed down into the lower ionosphere, conditions are quite different; it will be seen that the flow of current into the lower ionosphere controls the speed of the motion. At some depth the Ohmic field becomes important, and **E** is not perpendicular to **H**, and in considering the motion of the field the method of § 3.3 must be used. If the field is taken to move with the material in the equatorial F region, it must move with a similar motion everywhere, since the field must not be appreciably disturbed; the motion may be pictured as being resisted by the neutral particles in the lower ionosphere, and this explains how the speed is controlled. Alternatively, the motion in the equatorial F region may be regarded as that of a gravity-driven dynamo which drives a current along the lines of force into a load determined by the properties of the lower ionosphere; if a line of force is regarded as a cable, the performance of the cable deteriorates with depth, because σ_0 decreases while σ_1 increases. The condition that the magnetic field is not appreciably disturbed is expressed by putting $\mathbf{E} = -\nabla\phi$.

Since gravity provides the driving force, it is clear that the approximation (9.3) cannot be used, but a rough calculation can be made with the following simplified model. Let the lines of force be 'straightened out', so that $\mathbf{H}$ is uniform and n_n increases with the distance from the geomagnetic equator. Cartesian axes will be used with the x-axis pointing north, the y-axis west and the z-axis upwards; the model will be symmetrical with respect to the geomagnetic equatorial plane $x=0$. The model will be divided into two regions; in the 'dynamo region', defined by $x<a$, n_n is small and collisions will be completely neglected, while in the 'load region', $x>a$, the forces of gravity and the pressure gradient will be neglected and (9.7) and (9.8) will be used. The value of a is determined by the latitude at which a line of force has fallen say 200 km. below its height at the equator; this is about 10° and so $a \sim 1000$ km.

In the dynamo region, neglecting the acceleration of the convective motion, the forces are balanced according to

$$\mathbf{j} \wedge \mathbf{H}/c = 2\nabla p_e - \mu \mathbf{g}, \tag{9.31}$$

where μ refers to the charged particles only. Then

$$\frac{\partial j_y}{\partial y} + \frac{\partial j_z}{\partial z} = -\frac{gc}{H}\frac{\partial \mu}{\partial y}$$

and the vanishing of $\operatorname{div}\mathbf{j}$ together with the symmetry about the plane $x=0$ require

$$j_x = \frac{gc}{H}\int_0^x \frac{\partial \mu}{\partial y}\,dx, \tag{9.32}$$

which determines the current flowing into the load region.

In the load region (9.7) and (9.8) (with $\mathbf{u}_n=0$) become

$$\left.\begin{aligned} j_x &= -\sigma_0\,\partial\phi/\partial x, \\ j_y &= -\sigma_1\,\partial\phi/\partial y-\sigma_2\,\partial\phi/\partial z, \\ j_z &= -\sigma_1\,\partial\phi/\partial z+\sigma_2\,\partial\phi/\partial y, \end{aligned}\right\} \tag{9.33}$$

and from the last two the condition of continuity is

$$\frac{\partial j_x}{\partial x}=\frac{\partial}{\partial y}\left(\sigma_1\frac{\partial\phi}{\partial y}\right)+\frac{\partial}{\partial z}\left(\sigma_1\frac{\partial\phi}{\partial z}\right)+\left(\frac{\partial\phi}{\partial z}\frac{\partial}{\partial y}-\frac{\partial\phi}{\partial y}\frac{\partial}{\partial z}\right)\sigma_2. \tag{9.34}$$

This equation will now be grossly simplified by approximations which are partially justified by the result that the speed of the motion is proportional to its scale in the east-west direction. There must be some lower limit to this scale—for instance it cannot be less than the radius of the orbits of the positive ions, which is ~ 10 m.—but it is reasonable to suppose that the important term on the right-hand side of (9.34) is $\sigma_1\,\partial^2\phi/\partial y^2$. The problem is then linear and the Fourier component with $\phi\propto\cos ly$ will now be discussed. Then (9.34) becomes

$$\partial j_x/\partial x=-\sigma_1 l^2\phi \tag{9.35}$$

which must be solved in conjunction with (9.33).

Now, when $\nu_p\ll\Omega_p$, (9.10) shows that $\sigma_1\approx n_e e^2\nu_p/m_p\Omega_p^2$ and then $\sigma_1\sigma_0$ is independent of n_n. If the variation of n_e is neglected, $\sigma_1\sigma_0$ is then independent of x and the required solution of (9.33) and (9.35) is

$$\phi\propto\exp\left[-l\int(\sigma_1/\sigma_0)^{\frac{1}{2}}\,dx\right] \tag{9.36}$$

with
$$j_x=l(\sigma_0\sigma_1)^{\frac{1}{2}}\,\phi. \tag{9.37}$$

This approximation is not sound unless the exponent in (9.36) reaches a large value before ν_p becomes as large as Ω_p, and with the approximation used $(\sigma_1/\sigma_0)^{\frac{1}{2}}=0{\cdot}026\nu_p/\Omega_p$; consequently the approximation is poor unless l^{-1} is considerably less than 1 km., but may be used to obtain the order of magnitude of the speed of the motion.

In the dynamo region (9.1) approximates to (1.1) so that $u\sim cE/H$. Then from (9.37) and (9.32)

$$u\sim 0{\cdot}026\,alg\Omega_p^{-1}\,\delta n_e/n_e,$$

where δn_e is the perturbation in n_e in the irregularities. With $a \sim 1000$ km. and $l \sim 10$ km.$^{-1}$, $u \sim 30\delta n_e/n_e$ km./hr. and may be appreciable.

At present it is difficult to discern the effect of convective diffusion on the average properties of the layer, because their behaviour is complicated by other movements, but irregularities can be observed; at low geomagnetic latitudes 'spread echoes' set in rapidly after sunset.

9.5. The influence of the geomagnetic field on turbulence

It seems plausible that some of the irregularities in the ionosphere are caused by turbulent motion of the neutral molecules, and the

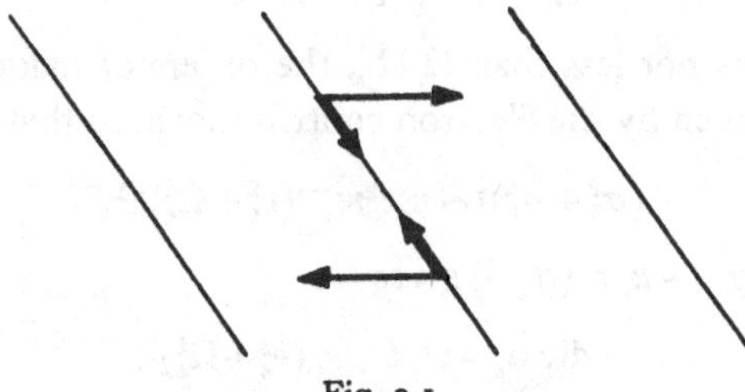

Fig. 9.1

effect of the magnetic field on such turbulence is discussed here. The effect on the motion of the neutral particles themselves can arise only through collisions with the charged particles, and, since $\nu_{n,p}$ is never much more than 10^{-3} sec.$^{-1}$, could only be important for very large eddies. The motion of the charged particles is driven by that of the neutral particles and is affected by the magnetic field.

Since it is variations in electron density which are observed, the important question is whether large variations of electron density can result from a motion in which n_n varies very little. It is easily seen that this occurs when $\nu_p \ll \Omega_p$, although the viscosity is then large and may well prevent turbulence from occurring. When $\nu_p \ll \Omega_p$ the motion of the charged particles is constrained to be almost parallel to $\mathbf{H}$, and the component of their velocity parallel to $\mathbf{H}$ is the same as that of the neutral particles, because it is still controlled by collisions; thus $\mathbf{u}_e \approx \mathbf{u}_p \approx (\mathbf{u}_n . \mathbf{h})\,\mathbf{h}$. An illustration of a motion which does not alter n_n is shown in fig. 9.1, where the sloping lines indicate the direction of **H**, the horizontal arrows represent $\mathbf{u}_n$, and the sloping arrows $\mathbf{u}_e$. Fig. 9.1 shows that the charged particles are being strongly compressed. When ν_p is not so small, the magnetic field still obstructs the motion of the charged particles across it and the effect still occurs, but is weaker. This is now treated quantitatively for values of ν_p substantially larger than Ω_p.

The partial pressure of the charged particles opposes the changes in n_e, but will be discussed later and (9.3) will be assumed to be valid. Then, if $\operatorname{div}\mathbf{u}_n$ vanishes,

$$\operatorname{div}\mathbf{u}_e \approx \operatorname{div}\mathbf{U} = (n_e m_p \nu_p c)^{-1} \operatorname{div}(\mathbf{j} \wedge \mathbf{H}). \tag{9.38}$$

Now, using (9.8)

$$\operatorname{div}(\mathbf{j} \wedge \mathbf{H}) \sim (\sigma_1^2 + \sigma_2^2)^{\frac{1}{2}} u_n H^2/ac,$$

and, if $\nu_e \nu_p$ is not less than $\Omega_e \Omega_p$, the orders of magnitude of σ_1 and σ_2 are given by the electron contributions, so that

$$(\sigma_1^2 + \sigma_2^2)^{\frac{1}{2}} \sim n_e e^2 m_e^{-1} (\nu_e^2 + \Omega_e^2)^{-1}.$$

Then writing $a \sim u_n t$, (9.38) gives

$$t \operatorname{div}\mathbf{u}_e \sim \Omega_p \Omega_e / \nu_p (\nu_e^2 + \Omega_e^2)^{\frac{1}{2}}. \tag{9.39}$$

This gives the order of magnitude of $\delta n_e/n_e$ and is seen to increase as n_n decreases. At 110 km. $\nu_p \sim 40\Omega_p$, so that $\delta n_e/n_e \sim 0{\cdot}025$. At greater heights (9.39) is not valid, because it was assumed that $\nu_e \nu_p > \Omega_e \Omega_p$, but symmetry between the electrons and positive ions suggests that the correct order of magnitude is obtained by replacing the ν_p in the denominator by $(\nu_p^2 + \Omega_p^2)^{\frac{1}{2}}$. Then, when $\nu_p < \Omega_p$, $\delta n_e \sim n_e$ as suggested by the illustration of fig. 9.1.

The partial pressure of the charged particles opposes the changes in n_e and makes a contribution to **U** of $-2(n_e m_p \nu_p)^{-1} \nabla p_e$. This is $\sim V_S^2 \delta n_e / a \nu_p n_e$, where V_S is the velocity of sound for the charged particles, but is of the same order of magnitude as that for the neutral particles. Further $\nu_p \sim \nu_{n,n}$ and so $V_S^2/\nu_p \sim$ the kinematic viscosity for the neutral particles and $V_S^2/a\nu_p \sim u_n/R$, where R is the Reynolds number for the turbulence. Then the contribution to δn_e due to the partial pressure gradient, which is $\sim V_S^2 \delta n_e / a \nu_p u_n$, is $\sim \delta n_e / R$; since $R \gg 1$ is a necessary condition for turbulence to persist, this contribution may be neglected.

9.6. Alfvén waves in the ionosphere

The propagation of Alfvén waves in the outer atmosphere was briefly discussed in § 8.5, particularly in connexion with the micropulsations which have periods of ~ 1 min. The propagation of these waves through the ionosphere needs to be discussed and it has also been suggested that Alfvén waves are useful for the description of ionospheric irregularities. Since collisions with the neutral particles are important here, the motion of the neutral particles must be considered; this itself results from collisions with the charged particles. Consider transverse waves, so that there is no disturbance in the pressure. Then, writing $\beta_{i,j}$ for $m_j \nu_{i,j}/(m_i + m_j)$ and considering a disturbance which varies like $e^{i\omega t}$,

$$i\omega' \mathbf{u}_n = \beta_{n,e} \mathbf{u}_e + \beta_{n,p} \mathbf{u}_p, \tag{9.40}$$

where

$$\omega' = \omega - i(\beta_{n,e} + \beta_{n,p}).$$

Now $\beta_{n,e} \ll \beta_{n,p}$ and (9.40) shows that, when $\omega \ll \beta_{n,p}$, the oscillations of the neutral particles are nearly the same as those of the positive ions, whereas, when $\omega \gg \beta_{n,p}$, they are smaller in amplitude and lag by nearly $\frac{1}{2}\pi$ in phase.

The dispersion equation will now be obtained for circularly polarized waves travelling in the direction of $\mathbf{H}_0$; the inertial terms and collisions of all kinds will be included, and no approximations will be made in the derivation. For circularly polarized waves the components are related as in $\mathbf{E} \wedge \mathbf{H}_0 = \pm iH_0 \mathbf{E}$, and the + sign corresponds to the direction of gyration of positive particles. Then the motion of the electrons is given by

$$i(\omega \pm \Omega_e)\mathbf{u}_e = -(e/m_e)\mathbf{E} + \beta_{e,n}(\mathbf{u}_n - \mathbf{u}_e) + \beta_{e,p}(\mathbf{u}_p - \mathbf{u}_e),$$

and there is a similar equation for the positive ions. Substituting for $\mathbf{u}_n$ from (9.40) gives

$$im_p \omega_p \mathbf{u}_p = -im_e \omega_e \mathbf{u}_e = e\mathbf{E} + m_e \beta'_{e,p}(\mathbf{u}_e - \mathbf{u}_p), \tag{9.41}$$

where

$$\omega_p = \omega \mp \Omega_p - i\omega\beta_{p,n}/\omega',$$

$$\omega_e = \omega \pm \Omega_e - i\omega\beta_{e,n}/\omega',$$

and

$$\beta'_{e,p} = \beta_{e,p} - i\beta_{e,n}\beta_{n,p}/\omega'.$$

Writing

$$\beta = (m_p \beta_{p,n} + m_e \beta_{e,n})/(m_p + m_e),$$

we note that

$$m_p \omega_p + m_e \omega_e = \omega(m_p + m_e)(1 - i\beta/\omega').$$

Then from (9.41)

$$\left(\frac{im_e m_p \omega_e \omega_p}{\omega(m_p+m_e)(1-i\beta/\omega')}+m_e\beta'_{e,p}\right)(\mathbf{u}_p-\mathbf{u}_e)=eE. \quad (9.42)$$

Now from Maxwell's equations, writing curl curl $\mathbf{E}=(\omega/V)^2\mathbf{E}$,

$$((c/V)^2-1)\,\omega\mathbf{E}=-4\pi i\mathbf{j}, \quad (9.43)$$

and, writing $V_A^2=H_0^2(4\pi ne(m_p+m_e))^{-1}$, the dispersion equation obtained from (9.42) and (9.43) is

$$V^2=-V_A^2(1-V^2/c^2)(\Omega_e\Omega_p)^{-1}\left\{\frac{\omega_e\omega_p}{1-i\beta/\omega'}-\frac{i\omega(m_p+m_e)\beta'_{e,p}}{m_p}\right\}. \quad (9.44)$$

This may be compared with the results of § 5.4; (5.13) is obtained from (9.44) by neglecting all collisions and neglecting ω/Ω_e and V/c. When collisions are included, but $\omega\to 0$, $\beta/\omega'\to in_n/n_e$ and so $V^2\to V_A^2/(1+n_n/n_e)$. It may also be verified that the Appleton–Hartree equation is obtained when $\omega\gg\Omega_p$. The approximation $\beta\approx\beta_{p,n}$ is always true and so, for $\omega\gg\Omega_p$, $\omega_p\approx\omega(1-i\beta/\omega')$. Also $\beta_{e,n}\beta_{n,p}\sim 10^{-11}n_e n_n$ sec.$^{-2}$ and this is less than Ω_p^2 in the ionosphere, whence, for $\omega\gg\Omega_p$, $\beta'_{e,p}\approx\beta_{e,p}$. Then the motion of the positive ions is unimportant, and the Appleton–Hartree treatment is valid.

Consider now the approximations valid for a period of 1 min. in the ionosphere. Since $\beta_{n,p}$ is never much greater than 10^{-3} sec.$^{-1}$, $\omega'\approx\omega$, and, since $\omega\ll\Omega_p$, $\omega_p\approx\mp\Omega_p-i\beta_{p,n}$ and $\omega_e\approx\pm\Omega_e-i\beta_{e,n}$. Then

$$\left|\frac{\omega_e\omega_p}{1-i\beta/\omega'}\right|\approx\left|\frac{(\Omega_e+i\beta_{e,n})(\Omega_p+i\beta_{p,n})}{1+i\beta_{p,n}/\omega}\right|>\omega\,|\,\Omega_e+i\beta_{e,n}\,|.$$

The other term in the curly brackets in (9.44) is then seen to be negligible by the following argument: $\beta_{e,p}\sim 100n_e T^{-\frac{3}{2}}$ sec.$^{-1}$, which is always small compared to Ω_e, so that $\omega\beta_{e,p}\ll\omega\Omega_e$, and, since $\beta_{n,p}\ll\omega$, $\beta_{e,n}\beta_{n,p}\ll\omega\beta_{e,n}$. It may also be verified that $V^2\ll c^2$ in the ionosphere and, reverting now to the notation ν_e and ν_p for $\beta_{e,n}$ and $\beta_{p,n}$, we have

$$V^2=V_{A^2}\frac{(1\mp i\nu_e/\Omega_e)(1\pm i\nu_p/\Omega_p)}{1-i\nu_p/\omega}$$

$$=\frac{3\,.\,10^{10}}{n_e}\frac{(1\mp i1{\cdot}5\,.\,10^{-15}n_n)(1\pm i2{\cdot}5\,.\,10^{-12}n_n)}{1-i4\,.\,10^{-4}n_n}\,(\text{km./sec.})^2,$$

for N_2^+ with $H = 0{\cdot}4$ gauss and $\omega = 0{\cdot}1$ sec.$^{-1}$. Writing k for the wave number this formula has been used to obtain estimates of $k^2 = 10^{-2}V^{-2}$ km.$^{-2}$ at various heights, the results being shown in table 4. This shows that $|k|^{-1}$ is everywhere greater than 100 km.

It must not be forgotten that waves of the sort considered here are reflected at the ground. The reflexion can be described by putting $\mathbf{j} = \sigma\mathbf{E}$ in (9.43); taking the resistivity of the earth's crust as 10^4 ohm cm. or 10^{13} e.m.u., we have $k^2 \sim \mathrm{i}\,10^{-3}$ km.$^{-2}$ inside the earth.

TABLE 4

Height (km.)	n_n (cm.$^{-3}$)	n_e (cm.$^{-3}$)	k^2 (km.$^{-2}$)
80	5.10^{14}	10^3	$\mp 3.10^{-7} - \mathrm{i}2\ .10^{-7}$
100	2.10^{13}	10^5	$\mp 5.10^{-5} - \mathrm{i}1{\cdot}5.10^{-6}$
120	10^{12}	10^5	$\mp 4.10^{-5} - \mathrm{i}1{\cdot}7.10^{-5}$
150	10^{11}	10^5	$\mp 3.10^{-6} - \mathrm{i}1{\cdot}1.10^{-5}$
200	10^{10}	10^5	$3.10^{-8} - \mathrm{i}1{\cdot}2.10^{-6}$
300	10^9	10^6	$3.10^{-7} - \mathrm{i}1{\cdot}2.10^{-6}$
400	10^8	2.10^5	$6.10^{-8} - \mathrm{i}2{\cdot}4.10^{-8}$

The simplest case will now be discussed in which the magnetic field is vertical; then, if there is no horizontal variation of $\mathbf{E}$,

$$\mathrm{d}^2E/\mathrm{d}z^2 = -k^2E, \tag{9.45}$$

with k^2 given by table 4. In the earth $E^{-1}\,\partial E/\partial z$ is $\sim 0{\cdot}02(1+\mathrm{i})$ km.$^{-1}$ and both E and $\partial E/\partial z$ are continuous at the surface. For a circularly polarized component let $q = E(\mathrm{d}E/\mathrm{d}z)^{-1} - z$; then from (9.45)

$$\mathrm{d}q/\mathrm{d}z = k^2(q+z)^2, \tag{9.46}$$

and, when $z = 0$, $q = 25(1-\mathrm{i})$ km. Now table 4 shows that

$$\int_0^{400} k^2\,\mathrm{d}z \sim \mp 2.10^{-3} - \mathrm{i}5.10^{-4}\,\text{km.}^{-1}, \quad \int_0^{400} k^2 z\,\mathrm{d}z \sim \mp 0{\cdot}2 - \mathrm{i}(0{\cdot}07)$$

and

$$\int_0^{400} k^2 z^2\,\mathrm{d}z \sim (4 \mp 20 - \mathrm{i}\,10)\,\text{km.};$$

consequently the change in q in going up to 400 km. is of the same order as the value of q at the ground. The amplitude reflexion coefficient R looking down from 400 km. is

$$|1 - \mathrm{i}k(q+z)|/|1 + \mathrm{i}k(q+z)|;$$

with $k \sim 2{\cdot}5.10^{-4}$ km.$^{-1}$, $|k(q+z)| \ll 1$, and hence $R \approx 1 + 2k\mathscr{I}(q)$. The rough integration performed above gives $\mathscr{I}(q) \approx -35 \pm 10$ km.

and hence $R \approx 0{\cdot}982 \pm 0{\cdot}005$. Consequently the earth, seen through the ionosphere, appears as a good reflector, but there is an appreciable difference in the reflection coefficients for the two polarizations.

To investigate the propagation of waves whose wavefronts are oriented obliquely to $\mathbf{H}_0$ it is necessary to consider the component of $\mathbf{j}$ parallel to $\mathbf{H}_0$. This is determined by the same expression as (9.42) with Ω_e and Ω_p omitted in the definition of ω_e and ω_p. Then, when $\nu_e \ll \Omega_e$, the factor on the left-hand side is much smaller than for the components perpendicular to $\mathbf{H}_0$ and it follows that $\mathbf{E}$ is nearly perpendicular to $\mathbf{H}_0$. Then, if χ is the angle of dip, the circularly polarized components may be written $E_\pm = E_N \sec\chi \mp iE_E$ where the suffices refer to north and east. If there is no horizontal variation, $\mathbf{j}$ is horizontal and $j_\pm = j_N \cos\chi \mp ij_E$. Then writing $k_\pm$ for the wave numbers given by (9.44) for the circularly polarized components, we have

$$\frac{d^2E_N}{dz^2} = -\tfrac{1}{2}\sec\chi((k_+^2+k_-^2)E_N\sec\chi - i(k_+^2-k_-^2)E_E), \quad (9.47)$$

$$\frac{d^2E_E}{dz^2} = -\tfrac{1}{2}((k_+^2+k_-^2)E_E + i(k_+^2-k_-^2)E_N\sec\chi). \quad (9.48)$$

The polarization ellipse now varies with height, but for moderate values of $\sec\chi$ the effect of the ionosphere is still not severe. When $\sec\chi$ is large, the expression in brackets on the right-hand side of (9.47) must be small and then (9.48) becomes

$$\frac{d^2E_E}{dz^2} \approx -\frac{2k_+^2k_-^2}{k_+^2+k_-^2}E_E,$$

and the effect of the E layer is then stronger, in a way similar to the equatorial enhancement of the tidal currents. At 100 km.,

$$2k_+^2k_-^2(k_+^2+k_-^2)^{-1} \approx -11{\cdot}7\,.\,10^{-3}\,\text{km.}^{-2}$$

and this is sufficient to make an order of magnitude change in q; considerable absorption therefore occurs.

APPENDIX

THE EFFECTIVE FREQUENCY OF COLLISIONS BETWEEN CHARGED PARTICLES

A rough explanation is given here of the logarithmic factor in the effective cross-section for collisions between charged particles. The collisions are intended to describe the effect of the Coulomb field of particles within the smoothing distance $n^{-\frac{1}{3}}$; this was the subdivision (2*a*) of § 2.4, and its effect is almost independent of the other parts of the field, which are here ignored. The magnitude of **E** at any point will generally be of the same order as that of the Coulomb field of the nearest particle. When a particle passes its nearest neighbour, their distance of closest approach being b, the field acting on the particle usually persists in the same direction for a time of order b/v, where v is the thermal velocity of the lighter particle. It will now be supposed that the particle making the collision has speed v; this is reasonable if the two particles are of the same kind or if it is the lighter particle (an electron) and the effective collision frequency for the heavier particle can then be obtained from the conservation of momentum. In the time b/v, a particle of speed v is deflected through an angle $\sim e^2/mv^2b$. If $b < e^2/mv^2$, we say that the velocity of the particle is completely randomized by the deflexion, but its velocity can also be completely randomized by a sucession of small deflexions. These small deflexions are in fact so numerous as to be the more important.

The cross-section for deflexions with b in the range δb is $2\pi b\,\delta b$ for values of b much less than the mean distance between particles $n^{-\frac{1}{3}}$. When b is not so small there is an appreciable probability that the particle at distance b is not the nearest neighbour, and consequently the cross-section is less than $2\pi b\,\delta b$. The cross-section will now be approximated as $2\pi b\,\delta b$ for $b < n^{-\frac{1}{3}}$ and zero for $b > n^{-\frac{1}{3}}$. The number of deflexions in time t with b in the range δb is $2\pi nvtb\,\delta b$. Successive deflexions are directed at random, so that the expected total deflexion after a time t is obtained by taking the square root of the sum of the squares of the individual deflexions. This can be written in integral form

$$\frac{e^2}{mv^2}\left(2\pi nvt\int_{e^2/mv^2}^{n^{-\frac{1}{3}}} \mathrm{d}b/b\right)^{\frac{1}{2}}.$$

The velocity of the particle is taken to become completely random when this deflexion = 1, or

$$t = t_1 = \left(\frac{mv^2}{e^2}\right)^2 (2\pi nv \log(n^{-\frac{1}{3}}mv^2/e^2))^{-1}.$$

The effect of small deflexions may therefore be represented as a contribution to the collision frequency of t_1^{-1}, and, including the effect of encounters with $b < e^2/mv^2$, the effective collision frequency has the form

$$\nu = \frac{\pi ne^4}{m^2v^3}(\alpha + 2\beta\log(n^{-\frac{1}{3}}mv^2/e^2))^{-1},$$

where α and β are not many times greater or less than 1. The result of Cowling's more rigorous calculation has this form.

Cohen, Spitzer and Routly (1950) have shown that thermal fluctuations in the charge density lead to correlation between the deflexions and that this can be accounted for approximately by replacing the cut-off distance $n^{-\frac{1}{3}}$ by the Debye distance $(kT/4\pi ne^2)^{\frac{1}{2}}$; this increases β by the factor $\frac{3}{2}$.

BIBLIOGRAPHY

GENERAL

ALFVÉN, H. (1950*a*). *Cosmical Electrodynamics* (Oxford University Press).

ALFVÉN, H. (1950*b*). Origin of solar magnetic fields. *Tellus*, **2**, 74.

ALFVÉN, H. (1954). *On the Origin of the Solar System* (Oxford University Press).

BABCOCK, H. W. and COWLING, T. G. (1953). General magnetic fields in the sun and stars. *Mon. Not. R. astr. Soc.* **113**, 357.

BIERMANN, L. (1952). Origin of magnetic fields in moving plasma. *Ann. Phys.* **10**, 413.

BLACKETT, P. M. S. (1947). The magnetic field of massive rotating bodies. *Nature, Lond.*, **159**, 658.

BOSTICK, W. H. (1956). Experimental study of ionised matter projected across a magnetic field. *Phys. Rev.* **104**, 292.

CHANDRASEKHAR, S. (1956). Axisymmetric magnetic fields and fluid motions. *Astrophys. J.* **124**, 232.

CHANDRASEKHAR, S. and FERMI, E. (1953*a*). Magnetic fields in spiral arms. *Astrophys. J.* **118**, 113.

COLE, G. H. A. (1956). Some aspects of magnetohydrodynamics. *Advanc. Phys.* **5**, 452.

COWLING, T. G. (1945). Conductivity of an ionised gas, with applications. *Proc. Roy. Soc.* A, **183**, 453.

DAVIS, L. and GREENSTEIN, J. L. (1951). The polarisation of starlight by aligned dust grains. *Astrophys. J.* **114**, 206.

ELLISON, M. A. (1955). *The Sun and its Influence* (Kegan Paul).

ELSASSER, W. M. (1954). Dimensional relations in magnetohydrodynamics. *Phys. Rev.* **95**, 1.

ELSASSER, W. M. (1956*a*). Hydromagnetism. A review. *Amer. J. Phys.* **24**, 85.

FERRARO, V. C. A. (1937). Non-uniform rotation of the sun and its magnetic field. *Mon. Not. R. astr. Soc.* **97**, 458.

GASIOROWICZ, S., NEUMANN, M. and RIDDELL, R. J. (1956). Dynamics of ionised media. *Phys. Rev.* **101**, 922.

LEHNERT, B. (1952). On the behaviour of an electrically conductive liquid in a magnetic field. *Ark. Fys.* **5**, 69.

LEHNERT, B. ed. (1958). Electromagnetic Phenomena in Cosmical Physics (Cambridge University Press).

LEHNERT, B. (1955). Magnetohydrodynamics on cosmical and laboratory scale. *K. tekn. Högsk. Avh.* no. 100.

LUNDQUIST, S. (1949). Experimental investigations of magneto-hydrodynamic waves. *Phys. Rev.* **76**, 1805.

LUNDQUIST, S. (1952). Studies in magnetohydrodynamics. *Ark. Fys.* **5**, 297.

LÜST, R. and SCHLÜTER, A. (1955). Angular momentum transport by magnetic fields. *Z. Astrophys.* **38**, 190.

ÖPIK, E. J. (1954). Interplanetary gas. *Z. Astrophys.* **35**, 43.

PIDDINGTON, J. H. (1955). The motion of ionised gas in magnetic, electric and mechanical fields. *Mon. Not. R. astr. Soc.* **114**, 651.

SCHLÜTER, A. (1952). Plasma in a magnetic field. *Ann. Phys.* **10**, 422.

SPITZER, L. (1954). Behaviour of matter in space. *Astrophys. J.* **120**, 1.

SPITZER, L. (1956). *The Physics of Fully Ionised Gases* (Interscience).

SYMPOSIUM (1955). *Gas Dynamics of Cosmic Clouds* (North Holland).

SYMPOSIUM IN LONDON (1955). (13 papers.) *Proc. Roy. Soc.* A, **233**, 289.

WRUBEL, M. H. (1952). The decay of a primeval stellar magnetic field. *Astrophys. J.* **116**, 291.

THEORY OF IONIZED GASES

CHAPMAN, S. (1954). The viscosity and thermal conductivity of a completely ionised gas. *Astrophys. J.* **120**, 151.

CHAPMAN, S. and COWLING, T. G. (1939). *Mathematical Theory of Non-uniform Gases* (Cambridge University Press, 2nd ed. 1952).

CHEW, G. F., GOLDBERGER, M. L. and LOW, F. E. (1956). The Boltzman equation and hydromagnetic equations. *Proc. Roy. Soc.* A, **236**, 112.

COHEN, R. S., SPITZER, L. and ROUTLY, P. McR. (1950). The electrical conductivity of an ionised gas. *Phys. Rev.* **80**, 230.

COWLING, T. G. (1956). The dissipation of magnetic energy in an ionised gas. *Mon. Not. R. astr. Soc.* **116**, 114.

GROSS, E. P. and KROOK, M. (1956). Model for collision processes in gases. *Phys. Rev.* **102**, 593.

LANDSHOFF, R. (1951). Convergence of the Chapman-Enskog method for a completely ionised gas. *Phys. Rev.* **82**, 442.

MASSEY, H. S. W. and BURHOP, E. H. S. (1952). *Electronic and Ionic Impact Phenomena* (Oxford University Press).

NEUFELD, J. and RITCHIE, R. H. (1955). Passage of charged particles through plasma. *Phys. Rev.* **98**, 1632.

OSTERBROCK, D. E. (1952). The electrical conductivity in the solar atmosphere. *Phys. Rev.* **87**, 468.

SPITZER, L. (1952). Equations of motion for an ideal plasma. *Astrophys. J.* **116**, 299.

SPITZER, L. and HARM, R. (1953). Transport phenomena in a completely ionised gas. *Phys. Rev.* **89**, 977.

WESTFOLD, K. C. (1953). The conduction current in an ionised gas. *Phil. Mag.* **44**, 712.

DYNAMO MECHANISMS

BONDI, H. and GOLD, T. (1950). On the generation of magnetism by fluid motion. *Mon. Not. R. astr. Soc.* **110**, 607.

COWLING, T. G. (1934). The magnetic field of sunspots. *Mon. Not. R. astr. Soc.* **94**, 39

DAVIS, L. (1956). Fluid self-excited dynamo. *Phys. Rev.* **102**, 939.

ELSASSER, W. M. (1956*b*). Hydromagnetic dynamo theory. *Rev. Mod. Phys.* **28**, 135.

McDONALD, K. L. (1954). Topology of steady current magnetic fields. *Amer. J. Phys.* **22**, 586.

PARKER, E. N. and KROOK, M. (1956). Diffusion and severing of magnetic lines of force. *Astrophys. J.* **124**, 214.

MAGNETOSTATICS

FERRARO, V. C. A. (1954). The equilibrium of magnetic stars. *Astrophys. J.* **119**, 407.

GJELLESTAD, G. (1954). On the elongation of a fluid sphere in a uniform external magnetic field. *Astrophys. J.* **120**, 172.

LÜST, R. and SCHLÜTER, A. (1954). Force-free magnetic fields. *Z. Astrophys.* **34**, 263.

MESTEL, L. (1956). The equilibrium of magnetic stars. *Mon. Not. R. astr. Soc.* **116**, 324.

PRENDERGAST, K. H. (1956). The equilibrium of a self-gravitating incompressible fluid sphere with a magnetic field. *Astrophys. J.* **123**, 498.

ALFVÉN WAVES AND OSCILLATIONS

AGOSTINELLI, C. (1954). Magnetohydrodynamic oscillations in a liquid cosmic mass. *Atti Accad. Torino* I, **89**, 68.

BABCOCK, H. W. (1951). The magnetically variable star HD 125248. *Astrophys. J.* **114**, 1.

BAÑOS, A. (1956). Normal modes characterising magneto-elastic plane waves. *Phys. Rev.* **104**, 300.

BURBIDGE, E. M. and BURBIDGE, G. R. (1955). Relative abundances and atmospheric conditions in the magnetic star HD 133029. *Astrophys. J.* **122**, 396.

CHANDRASEKHAR, S. (1956). Hydromagnetic oscillations of a fluid sphere with internal motions. *Astrophys. J.* **124**, 571.

COWLING, T. G. (1952). The oscillation theory of magnetic variable stars. *Mon. Not. R. astr. Soc.* **112**, 527.

DUNGEY, J. W. (1954). Attenuation of Alfvén waves. *J. geophys. Res.* **59**, 323.

GINZBURG, V. L. (1951). Magnetohydrodynamic waves in gases. *Zh. Eksper. Teor. Fiz.* **21**, 788.

HINES, C. O. (1953). Generalised magnetohydrodynamic formulae. *Proc. Camb. Phil. Soc.* **49**, 299.

LEHNERT, B. (1955). Magnetohydrodynamic waves under the action of the Coriolis force. *Astrophys. J.* **121**, 481.

LOUGHHEAD, R. E. (1955). Eigen oscillations of compressible ionised fluids. *Aust. J. Phys.* **8**, 416.

LYTTKENS, E. (1954). On the radial pulsations of an infinite cylinder with a magnetic field parallel to its axis. *Astrophys. J.* **119**, 413.

WALEN, C. (1949). *On the Vibratory Rotation of the Sun* (Stockholm: Henrik Lindstahls Bokhandel).

STABILITY

BERNSTEIN, I. B., FRIEMAN, E. A., KRUSKAL, M. D. and KULSRUD, R. M. (1958). An energy principle for hydromagnetic stability problems. *Proc. Roy. Soc.* A, **244**, 17.

CHANDRASEKHAR, S. (1952). The inhibition of convection by a magnetic field. *Phil. Mag.* **43**, 501.

CHANDRASEKHAR, S. (1953*a*). Problems of stability in hydrodynamics and hydromagnetics. *Mon. Not. R. astr. Soc.* **113**, 667.

CHANDRASEKHAR, S. (1953*b*). The instability of a layer of fluid heated below and subject to Coriolis forces. *Proc. Roy. Soc.* A, **217**, 306.

CHANDRASEKHAR, S. and FERMI, E. (1953*b*). Problems of gravitational stability in the presence of a magnetic field. *Astrophys. J.* **118**, 116.

DUNGEY, J. W. and LOUGHHEAD, R. E. (1954). Twisted magnetic fields in conducting fluids. *Aust. J. Phys.* **7**, 5.

HAIN, K., LÜST, R. and SCHLÜTER, A. (1957). Zur Stabilität eines Plasmas. *Z. Naturforschg.* **12a**, 833.

KRUSKAL, M. and SCHWARZSCHILD, M. (1954). Some instabilities of a completely ionised gas. *Proc. Roy. Soc.* A, **223**, 348.

LOCK, R. C. (1955). The stability of the flow between parallel planes under a transverse magnetic field. *Proc. Roy. Soc.* A, **233**, 105.

LOUGHHEAD, R. E. (1955). Hydromagnetic stability of a current layer. *Aust. J. Phys.* **8**, 319.

LUNDQUIST, S. (1951). The stability of magnetohydrostatic fields. *Phys. Rev.* **83**, 307.

STUART, J. T. (1954). The stability of viscous flow between parallel planes in the presence of a coplanar magnetic field. *Proc. Roy. Soc.* A, **221**, 189.

TANDBERG-HANSSEN, E. (1953). The influence of a magnetic field on the rotational stability of stars. *Ann. Astrophys.* **16**, 206.

TAYLER, R. J. (1957). Hydromagnetic instabilities. *Proc. Phys. Soc.* B, **70**, 31.

THOMPSON, W. B. (1951). Thermal convection in a magnetic field. *Phil. Mag.* **42**, 1417.

TURBULENCE

BATCHELOR, G. K. (1950). The spontaneous magnetic field in a conducting liquid in turbulent motion. *Proc. Roy. Soc.* A, **201**, 405.

CHANDRASEKHAR, S. (1951). The invariant theory of isotropic turbulence in magneto-hydrodynamics. *Proc. Roy. Soc.* A, **207**, 301.

CHANDRASEKHAR, S. (1955). Hydromagnetic turbulence. *Proc. Roy. Soc.* A, **233**, 322.

KRYZWOBLOCKI, M. Z. E. (1952). The decay of isotropic turbulence in magnetohydrodynamics. *J. Phys. Soc. Japan*, **7**, 511.

KULSRUD, R. M. (1955). Effect of magnetic fields on generation of noise by isotropic turbulence. *Astrophys. J.* **121**, 461.

LEHNERT, B. (1955). The decay of magneto-turbulence. *Quart. appl. Math.* **12**, 321.

SCHILLING, P. O. and LOCHTE-HOLTGREVEN, W. (1954). Magnetic fields in turbulent streaming plasmas. *Z. Naturf.* **99**, 520.

SWEET, P. A. (1950). The effect of turbulence on a magnetic field. *Mon. Not. R. astr. Soc.* **110**, 69.

SHOCK WAVES

DE HOFFMANN, F. and TELLER, E. (1950). Magnetohydrodynamic shocks. *Phys. Rev.* **80**, 692.

HELFER, L. (1953). Magnetohydrodynamic shock waves. *Astrophys. J.* **117**, 177.

LÜST, R. (1955). Stationary magnetohydrodynamic shock waves of arbitrary strength. *Z. Naturf.* **10a**, 125.

MARSHALL, W. (1955). The structure of magnetohydrodynamic shock waves. *Proc. Roy. Soc.* A, **233**, 367.

SEN, H. K. (1956). Structure of a magnetohydrodynamic shock wave in a plasma of infinite conductivity. *Phys. Rev.* **102**, 5.

COSMIC RAYS AND ACCELERATING MECHANISMS

ALPHER, R. A. (1950). Theoretical geomagnetic effects in cosmic radiation. *J. geophys. Res.* **55**, 437.

BIERMANN, L. (1953). Origin and propagation of cosmic rays. *Ann. Rev. Nuclear Sci.* **2**, 335.

BIERMANN, L. and SCHLÜTER, A. (1951). Cosmic radiation and cosmic magnetic fields. *Phys. Rev.* **82**, 863.

FAN, C. Y. (1956). Origin of cosmic radiations. *Phys. Rev.* **101**, 314.

FERMI, E. (1954). Galactic magnetic fields and the origin of cosmic radiation. *Astrophys. J.* **119**, 1.

GIOVANELLI, R. G. (1947). Electric phenomena associated with sunspots. *Mon. Not. R. astr. Soc.* **107**, 338.

KAPLAN, S. A. (1955). Theory of the acceleration of particles by isotropic turbulent fields. *Zh. Eksper. Teor. Fiz.* **29**, 406.

KAPLON, M. F., PETERS, B., REYNOLDS, H. L. and RITSON, D. M. (1952). The energy spectrum of primary cosmic radiation. *Phys. Rev.* **85**, 295.

LEPRINCE-RINGUET, L. (1950). *Cosmic Rays* (Constable).

MORRISON, P., OLBERT, S. and ROSSI, B. (1954). Origin of cosmic rays. *Phys. Rev.* **94**, 440.

PARKER, E. N. (1955). Hydromagnetic waves and the acceleration of cosmic rays. *Phys. Rev.* **99**, 241.

SWANN, W. F. G. (1933). A mechanism of acquirement of cosmic-ray energies by electrons. *Phys. Rev.* **43**, 217.

TELLER, E. (1954). Theory of origin of cosmic rays. *Rep. Progr. Phys.* **17**, 154.

THOMPSON, W. B. (1955). On the acceleration of particles by magnetohydrodynamic waves. *Proc. Roy. Soc.* A, **233**, 402.

THE SUN

ALLEN, C. W. (1954). The physical condition of the solar corona. *Rep. Progr. Phys.* **17**, 135.

ANDERSON, C. N. (1954). Notes on the sunspot cycle. *J. geophys. Res.* **59**, 455.

ATHAY, R. G. and THOMAS, R. N. (1956). Some characteristics of a chromospheric model. *Astrophys. J.* **123**, 309.

BABCOCK, H. W. and BABCOCK, H. D. (1955). The sun's magnetic field. *Astrophys. J.* **121**, 349.

DISCUSSION (1957). The event of February 23, 1956. *Observatory*, **77**, 88.

EDDINGTON, Sir A. S. (1942). Conditions in the hydrogen convection zone. *Mon. Not. R. astr. Soc.* **102**, 154.

FIROR, J. W., SIMPSON, J. A. and TRIEMAN, S. B. (1954). Production of cosmic radiation at the sun. *Phys. Rev.* **95**, 1015.

GLEISSBERG, W. (1953). The 80 year sunspot cycle in ancient auroral observations. *Naturwissenschaften*, **40**, 336.

KAHN, F. D. (1950). Streams of corpuscles. *Mon. Not. R. astr. Soc.* **110**, 483.

HOYLE, F. (1949). *Some Recent Researches in Solar Physics* (Cambridge Monographs on Physics).

KUIPER, G. ed. (1953). *The Sun* (Chicago).

LIPPINCOTT, S. L. (1955). Changes in chromospheric structure. *Ann. Astrophys.* **18**, 113.

MACRIS, C. (1955). Variation of photospheric granules. *Observatory*, **75**, 122.

MICHARD, R. (1953). The photosphere and sunspots. *Ann. Astrophys.* **16**, 217.

PARKER, E. N. (1955). Formation of sunspots. *Astrophys. J.* **121**, 491.

PIDDINGTON, J. H. (1956). Solar atmospheric heating by hydromagnetic waves. *Mon. Not. R. astr. Soc.* **116**, 314.

SIEDENTOPF, H., BEHR, A. and ELSASSER, H. (1953). Photoelectric observations of the zodiacal light. *Nature, Lond.*, **171**, 1066.

SWEET, P. A. (1955). 'The structure of sunspots.' *Vistas in Astronomy* (ed. Beer, Pergamon), p. 675.

VAN DE HULST, H. C. (1950). The electron density of the solar corona. *Bull. Astr. Insts. Netherlds*, **11**, 135.

WALDMEIER, M. (1950). Statistics of green coronal rays. *Z. Astrophys.* **27**, 237.

WALDMEIER, M. (1952). Variability of the coronal temperature. *Z. Astrophys.* **30**, 137.

WALDMEIER, M. (1954). Simultaneous determination of electron density and temperature in the solar corona. *Z. Astrophys.* **35**, 95.

WILD, J. P., MURRAY, J. D. and ROWE, W. C. (1954). Harmonics in the spectra of solar radio disturbances. *Aust. J. Phys.* **7**, 439.

MAGNETIC VARIATIONS

ASHOUR, A. A. and PRICE, A. T. (1948). The induction of currents in a non-uniform ionosphere. *Proc. Roy. Soc.* A, **195**, 198.

BEISER, A. (1955). On an interplanetary magnetic field. *J. geophys. Res.* **60**, 155.

CHAPMAN, S. (1951). *The Earth's Magnetism* (Methuen Monographs on Physical Subjects).

CHAPMAN, S. and BARTELS, J. (1940). *Geomagnetism* (Oxford).

DUNGEY, J. W. (1955). Electrodynamics of the outer atmosphere. *Proc. Ionosphere Conf.* (*Phys. Soc. London*), p. 229.

FERRARO, V. C. A. (1946). The radial stability of the geomagnetic ring-current. *Terr. Magn. atmos. Elect.* **51**, 547.

FERRARO, V. C. A. (1952). Theory of a plane model. *J. geophys. Res.* **57**, 15.

FESENKOV, V. G. (1950). The gas tail of the earth. *Astron. Zh.* **27**, 2. (Translation by Hope, D. R. B., Canada.)

FORBUSH, S. E. and VESTINE, E. H. (1955). Daytime enhancement of sudden commencements. *J. geophys. Res.* **60**, 299.

GRENET, G. (1955). Explanation of polar magnetic perturbations. *C.R. Acad. Sci., Paris*, **240**, 448.

HOLMBERG, E. R. R. (1953). Rapid periodic fluctuations of the geomagnetic field. *Mon. Not. R. astr. Soc. geophys. Suppl.* **6**, 467.

HOPE, E. R. (1956). Spiral patterning of solar corpuscular precipitation. *Nature, Lond.*, **177**, 571.

JACOBS, J. A. and OBAYASHI, T. (1956). The diurnal behaviour of sudden commencements. *Canad. J. Phys.* **34**, 876.

KATO, Y. (1951). A new theory of magnetic storms. *Rep. Ionos. Res. Japan*, **5**, 75.

KATO, Y. and AKASOFU, S-I. (1956). Relations between micropulsations and solar UM regions. *J. atmos. terr. Phys.* **9**, 352.

MEEK, J. H. (1955). The location and shape of the auroral zone. *J. atmos. terr. Phys.* **6**, 313.

NEWTON, H. W. and MILSOM, A. S. (1954). Distribution of geomagnetic storms in the sunspot cycle. *J. geophys. Res.* **59**, 203.

PARKER, E. N. (1956). On the geomagnetic storm effect. *J. geophys. Res.* **61**, 625.

SIMPSON J. A. (1955). The cosmic radiation and solar-terrestrial relationships. *Ann. Geophys.* **11**, 305.

SIMPSON, J. A., FENTON, K. B., KATZMAN, J. and ROSE, D. C. (1956). Effective geomagnetic equator for cosmic radiation. *Phys. Rev.* **102**, 1648.

TREIMANN, S. B. (1953). Effect of equatorial ring current on cosmic ray intensity. *Phys. Rev.* **89**, 130.

TROITSKAYA, V. A. (1953). Two types of oscillation of the earth's electromagnetic field. *Dokl. Akad. Nauk. SSSR*, **93**, 261.

AURORAE

ARMSTRONG, E. B. and DELGARNO, A. eds. (1956). *The Airglow and the Aurorae* (Pergamon).

BOOKER, H. G., GARTLEIN, C. W. and NICHOLS, B. (1955). Radio reflections from the aurora. *J. geophys. Res.* **60**, 1.

FERRARO, V. C. A. (1953). The aurorae. *Advanc. Phys.* **2**, 265.

HARANG, L. (1951). *The Aurorae* (Chapman and Hall, International Astrophysics Series).

HULBURT, E. O. (1928). Origin of the aurora. *Phys. Rev.* **31**, 1038.

JACKA, F. (1954). Variations of intensity of the aurora. *Aust. J. Phys.* **7**, 477.

MEEK, J. H. and MCNAMARA, A. G. (1954). Magnetic disturbances, sporadic *E* and radio echoes associated with the aurora. *Canad. J. Phys.* **32**, 326.

MEINEL, A. B. (1951). The spectrum of the airglow and the aurora. *Rep. Progr. Phys.* **14**, 121.

MEINEL, A. B. (1954). The morphology of the aurora. *Proc. Nat. Acad. Sci., Wash.*, **40**, 943.

MEINEL, A. B. and SCHULTE, D. H. (1953). A note on auroral motions. *Astrophys. J.* **117**, 454.

MEREDITH, L. H., GOTTLIEB, M. B. and VAN ALLEN, J. A. (1955). Direct detection of soft radiation above 50 km. in the auroral zone. *Phys. Rev.* **97**, 201.

STØRMER, C. (1955). *The Polar Aurora* (Oxford University Press).

SUGIURA, M., TAZIMA, M. and NAGATA, T. (1952). Anomalous ionization in the upper atmosphere over the auroral zone during magnetic storms. *Rep. Ionos. Res. Japan*, **6**, 147.

THE IONOSPHERE

APPLETON, Sir E. (1953). Storm phenomena in the ionosphere. *Arch. elektr. Übertr.* **7**, 271.

BOOKER, H. G. (1954). Morphology of ionospheric storms. *Proc. Nat. Acad. Sci., Wash.*, **40**, 931.

BRIGGS, B. H. and SPENCER, M. (1954). Horizontal movements in the ionosphere. *Rep. Progr. Phys.* **17**, 245.

CLEMMOW, P. C., JOHNSON, M. A. and WEEKES, K. (1955). Motion of a cylindrical irregularity in an ionised medium. *Proc. Ionosphere Conf. (Phys. Soc. London)*, p. 136.

DUNGEY, J. W. (1956). Convective diffusion in the equatorial F region. *J. atmos. terr. Phys.* **9**, 304.

FEJER, J. A. (1953). Semidiurnal currents in the ionosphere. *J. atmos. terr. Phys.* **4**, 184.

JOHNSON, M. H. and HULBURT, E. O. (1950). Diffusion in the ionosphere. *Phys. Rev.* **79**, 802.

KATO, S. (1956). Wind systems in the E region deduced from the dynamo theory. *J. Geomagn. Geoelect. Kyoto*, **8**, 24.

LEWIS, R. P. and MCINTOSH, D. H. (1953). Geomagnetic and ionospheric relationships. *J. atmos. terr. Phys.* **4**, 44.

MAPLE, E., BOWEN, W. A. and SINGER, S. F. (1951). Evidence for ionospheric currents from rocket experiments. *J. geophys. Res.* **56**, 265.

MARTYN, D. F. (1953). Ionospheric variations associated with magnetic disturbances. *Proc. Roy. Soc.* A, **218**, 1.

NAGATA, T. (1950). The auroral zone current. *Rep. Ionos. Res. Japan*, **4**, 87.

NICOLET, M. (1953). The collision frequency of electrons in the ionosphere. *J. atmos. terr. Phys.* **3**, 200.

STOREY, L. R. O. (1953). An investigation of whistling atmospherics. *Phil. Trans.* **246**, 113.

WEEKES, K. (1954). The physical state of the upper atmosphere. *Quart. J. R. Met. Soc.* **80**, 2.

WILKES, M. V. (1949). *Oscillations of the Earth's Atmosphere* (Cambridge monographs on Physics).